Multimodal Agents for Ageing and Multicultural Societies

Juliana Miehle · Wolfgang Minker ·
Elisabeth André · Koichiro Yoshino
Editors

Multimodal Agents for Ageing and Multicultural Societies

Communications of NII Shonan Meetings

Springer

Editors
Juliana Miehle
Institute of Communications Engineering
University of Ulm
Ulm, Germany

Wolfgang Minker
Institute of Communications Engineering
University of Ulm
Ulm, Germany

Elisabeth André
Human-Centered Multimedia
University of Augsburg
Augsburg, Germany

Koichiro Yoshino
Nara Institute of Science and Technology
Ikoma, Japan

ISBN 978-981-16-3478-9 ISBN 978-981-16-3476-5 (eBook)
https://doi.org/10.1007/978-981-16-3476-5

This Springer imprint is published by the registered company Springer Nature Singapore Pte Ltd.
The registered company address is: 152 Beach Road, #21-01/04 Gateway East, Singapore 189721, Singapore

Preface

Nowadays, intelligent agents are omnipresent and used by almost all generations. Furthermore, we live in a globally mobile society in which populations of widely different cultural backgrounds live and work together. The number of people who leave their ancestral cultural environment and move to countries with different culture and language is increasing. This is not free of challenges. Especially in the case of care, migrants often face two issues: (i) not being able to speak the language and not being acquainted with the culture of the resident country, and (ii) being unfamiliar with the care and health administrations of the country. As a consequence, e.g. elderly migrants in care homes suffer from social exclusion, with their relatives also struggling with getting the right information and interacting with the administration, migrants at home are often reluctant to go to see the doctor in case of health issues, a tendency that is often further aggravated by cultural matters. Migrant temporary care workers face the problem of isolation and deficient communication.

We believe that time is ripe to gather experts with a background in assistive technologies for elderly care, culture-aware computing, multimodal dialogue, social robotics and synthetic agents to learn about their insights and approaches on how to meet the resulting societal challenges. The overall objective of the Shonan meeting entitled "Multimodal Agents for Ageing and Multicultural Societies" has been to explore and discuss theories and technologies for the development of socially competent and culture-aware embodied conversational agents for elderly care. The results of the discussions will help to foster the understanding of the emerging field and the identification of promising approaches from a variety of disciplines. In order to share this knowledge, the results of thc meeting are published within this book.

Chapter 1 presents a vision of an intelligent agent to illustrate the current challenges for the design and development of adaptive systems. These are analysed following the what-which-how-then model to cover all aspects including which features to adapt, what to adapt to, when to adapt and how to adapt. A special emphasis is given to multimodal conversational interaction. It becomes clear that there is no single system behaviour that fits all users as contemporary societies are comprised of individuals very diverse in terms of culture, status, gender and age. Multicultural and ageing societies demand *adaptive* interactive systems with the ability to learn about and from their users and adjust their behaviour accordingly.

Chapter 2 examines how notions of *trust* and *empathy* may be applied to human-robot interaction and how it can be used to create the next generation of emphatic agents, which address some of the pressing issues in multicultural ageing societies. In order to do so, some examples of possible future services are given. The authors discuss how existing descriptive models of trust can be operationalized so as to allow robot agents to elicit trust in the human partners. Another topic is to determine how the robot agents can detect and alter trust levels to have transparency to be included in the interaction design. With regard to future computational models of empathy, fundamental challenges are formulated, and psychological models of empathy are discussed for their appliance in computational models and their use of interactive empathic agents.

Chapter 3 discusses multimodal machine learning as an approach to enable more effective and robust modelling technologies and to develop socially competent and culture-aware embodied conversational agents for elderly care. With the goal of better understanding and modelling the behaviour of ageing individuals, four key challenges are identified: (i) *multimodal*, this modelling task includes multiple relevant modalities which need to be represented, aligned and fused; (ii) *high variability*, this modelling problem expresses high variability given the many social contexts, large space of actions and the possible physical or cognitive impairment; (iii) *sparse and noisy resources*, this modelling challenge addresses unreliable sensory data and the limitation and sparseness of resources that are specific for the special user group of ageing individuals; and (iv) *concept drift*, where two types of drift were identified, namely on the group level (as the target group of usage is not fully known at the moment of development of according interfaces given that it is yet to age) and the individual level (given that ageing may lead to drifting behaviour and interaction preferences throughout the ongoing ageing effect).

Chapter 4 explores the challenges associated with real-world field tests and deployments. The authors review some of the existing methods and their suitability for studies with multimodal and multicultural agents. Moreover, ethical concerns related to field studies with sometimes vulnerable users are investigated. Moving out of the laboratory poses a number of challenges that are not present in controlled settings and requires to thoroughly prepare for the unpredictability of the field. Fortunately, it is not necessary to start from scratch but some of the methods developed in the human-computer interaction community for studies outside the lab can be re-uses.

Chapter 5 gives a short introduction to Socio-cognitive Language Processing which describes the idea of coping with everyday language (including slang and multi-lingual phrases and cultural aspects), irony, sarcasm, humour, paralinguistic information such as the physical and mental state and traits of the dialogue partner (e.g. affect, age groups, personality dimensions), and social aspects. Additionally, multimodal aspects such as facial expressions, gestures or bodily behaviour should ideally be included in the analysis where possible. These aspects may render future dialogue systems more 'chatty' by not only appearing natural but also by being truly emotionally and socially competent, ideally leading to a more symmetrical dialogue. To do this, the computer should be enabled to experience or at least better understand emotions and personality so that that they have ‘a feel’ for concepts like having a ‘need for humour’, an ‘increase of familiarity’.

We wish to thank the National Institute of Informatics and the NII Shonan Meeting Center for the excellent organisation of the meeting.

Ulm, Germany — Juliana Miehle
Ulm, Germany — Wolfgang Minker
Augsburg, Germany — Elisabeth André
Nara, Japan — Koichiro Yoshino
February 2021

Contents

Chapter 1
Adaptive Systems for Multicultural and Ageing Societies

Zoraida Callejas, Birgit Lugrin, Jean-Claude Martin, Michael F. McTear, and Juliana Miehle

1.1 Vision/Use Case

Ruth and Daniel are both 78 years old, they have been married for 50 years and live together in their house in a quiet neighbourhood on the outskirts of the city. Although they are retired now, they have very busy agendas, especially since their daughter Eva has a child, Alex.

Ruth enjoys cycling and taking dancing lessons with her friends. She also does voluntary work at the public library two days a week. She has always been very active, so she tries to do as much exercise as possible in her daily life, walking or cycling to her destinations when possible, using the stairs, and making plans with her friends that include some type of mild physical activity (e.g. walking outdoors). Unfortunately, she is losing her vision as she gets older.

Daniel used to be very sporty, but his joints ache now due to arthritis and he can only walk short distances. He takes care of a little garden and enjoys cooking the

Z. Callejas (✉)
University of Granada, Granada, Spain
e-mail: zoraida@ugr.es

B. Lugrin
University of Würzburg, Würzburg, Germany
e-mail: birgit.lugrin@uni-wuerzburg.de

J.-C. Martin
Laboratoire Interdisciplinaire des Sciences du Numerique, CNRS, Universite Paris-Saclay, Paris, France
e-mail: jean-claude.martin@limsi.fr

M. F. McTear
Ulster University, Belfast, UK
e-mail: mf.mctear@ulster.ac.uk

J. Miehle
University of Ulm, Ulm, Germany
e-mail: juliana.miehle@uni-ulm.de

J. Miehle et al. (eds.), *Multimodal Agents for Ageing and Multicultural Societies*,
https://doi.org/10.1007/978-981-16-3476-5_1

vegetables that he grows there in order to keep his blood pressure and sugar levels at a healthy level, as he had problems before. He also meets with friends at the local bar to watch sport together as often as he can, especially when his favourite team is playing. Every now and then he gives lessons to their neighbour's children who need help with maths at school.

Apart from their regular activities, they have frequent appointments with doctors to take care of their conditions (blood pressure, sugar level, sight...) and they take turns at taking their grandchild to school and every now and then they babysit at home or at their daughter's when she has to stay late at work.

Eva lives near her parents, approximately 30 min walk to Alex's school. She usually takes him to their house in the morning and goes to work, as school starts later. She is passionate about new technologies and has always been very active. She holds a responsible position within her company, so her agenda is always full, especially since her son was born.

Coordinating the agendas in this family is not easy, that is why Eva suggested to use an intelligent reminder agent (Rob). The system is connected to a smart speaker and a screen in Ruth and Daniel's kitchen, a smart speaker and screen in Eva's living room and to all of their smartphones. As Eva regularly uses her smartphone hands-free in her car, the system is also in some sense connected to Eva's car. The system has conversational capabilities, so it is possible to engage in a dialogue with it. This is very convenient for Ruth who cannot operate devices with tiny fonts and for Daniel who is usually in the kitchen. Eva is always on the run, so she can check her agenda while she is driving to work or tidying Alex's mess in the living room.

In addition, the system is able to adapt to everybody's needs, both permanent (e.g. considering Daniel's restrictions for walking) and transient (e.g. a broken leg). The next sections show some example interactions.

1.1.1 Example 1. Same Appointment, Different Contexts

Daniel has an appointment with the doctor at 17:00. Rob knows that the distance to the doctor is too long for Daniel to walk. It takes 20 min by taxi given the current traffic.

16:00

Rob: Hi Daniel, you have an appointment with Dr. M. at 17:00, shall I call a taxi for 16:30?

Daniel: Yes, please.

16:20

Rob: Your taxi will arrive in 10 min.

Ruth has an appointment with the doctor at 17:00. It takes 30 min by bike. This time, Rob reminds a little earlier, as it usually takes longer for Ruth to get ready to go out.

15:30

Rob: Hi Ruth, you have an appointment with Dr. M. at 17:00, if you are taking the bike you should leave in an hour.

Ruth has an appointment with the doctor at 17:00 again, this time there is a very heavy storm.

15:30

Rob: Hi Ruth, you have an appointment with Dr. M. at 17:00, it will be raining heavily, will you consider taking a taxi?

Ruth: Yes, that will be better.

Rob: Shall I order it for 16:15? It may take longer with the rain.

Ruth: Sure.

16:05

Rob: Your taxi will arrive in 10 min.

1.1.2 Example 2. Synchronizing Agendas

Eva is busy in the office, she wants to arrange a meeting for Tuesday, but she will need her parents to take care of Alex meanwhile. She does not have time to call them, so she asks Rob to attempt to synchronize agendas.

Eva on the phone

Eva: I have to arrange a meeting next Tuesday. Can somebody take care of Alex?

Rob: Ruth will be on a trip and Daniel is busy from 10:00 to 12:00.

Rob does not disclose why Daniel is busy unless Daniel specifies that it is a public task when creating the appointment.

Eva: OK, check if Daniel can manage from 13:00 to 17:00.

At Daniel's kitchen with the smart speaker

Rob: Hi Daniel, Eva is asking for babysitting.

Daniel: When?

Rob: Next Tuesday from 13:00 to 17:00. You are giving maths lessons earlier from 10:00 to 12:00.

Daniel: OK, schedule babysitting.

Rob: OK, I have scheduled babysitting from 13:00 to 17:00 on Tuesday 3rd November.

Rob knows that it should not interrupt Eva at work with phone calls, so it uses text messages to communicate with her.

Text chat with Eva

Rob: Babysitting scheduled for Daniel next Tuesday from 13:00 to 17:00.

Eva: Ok, schedule a meeting with the development team next Tuesday at 14:00 in the blue room.

Rob: Done.

1.1.3 Example 3. Implicit Appointments

Rob knows all the tasks in the family agenda, some of them are explicit (e.g. an appointment with the doctor), but some are not. For example, Rob knows that Alex finishes school at 13:30 and somebody must pick him up every day. Usually, Eva does it, so it is an implicit appointment in her agenda.

However, today Eva has scheduled a meeting from 12:30 to 13:00. As it takes her a long time to go from work to Alex's school, Rob infers that she cannot make it on time, this means that the implicit appointment for picking Alex from school falls today to Ruth or Daniel. Rob checks both agendas and sees that Daniel has a visit to the dentist, so it decides that it is Ruth's task to take Alex home today.

It is 08:30 and Ruth enters the kitchen to have breakfast.

Rob recognizes Ruth's face

Rob: Good morning Ruth!

Ruth: Good morning!

Rob: Shall I remind you of your complete agenda or just the urgent appointments?

Rob knows that it is good for Ruth to try to remember the regular appointments (for example, that on Mondays she goes to the library), so it does not remind her about them by default. However, there are appointments that are non-recurrent or that may be urgent and which are important to remind her about.

Ruth: Just the urgent ones this time.

Rob: OK. Today Eva has a late meeting, so you must pick Alex up from school. Daniel cannot do it this time.

Ruth: OK, got it. Anything else?

Rob: No, just the usual activities, shall I remind you of them?

Ruth: That's not necessary, thank you!

In Eva's phone:

Rob: As you have the meeting with the design team until 13:00, Ruth will pick Alex up from school today.

It is 12:30, Rob knows that Ruth prefers cycling and walking to places, so it usually reminds her with enough advance notice, as it takes her longer to reach places compared to taking public transport (as Daniel usually does) or driving (as Eva does). This time she is going to Alex's school, so Rob knows it will take her 30 min to walk and 15 min to cycle there.

From her smart phone, Rob knows Ruth is in the garden, so it phones her.

Ruth: Hi Rob, what's up?

Rob: It is 12:30, if you walk to Alex's school you will have to leave in 30 min.

Ruth: Thank you, I think I will take the bike.

Rob: Then you will have to leave at 13:15 at the latest. Shall I remind you 15 min beforehand?

Ruth: OK, thanks!

It is 13:15

Rob: Ruth, it is time to take the bike and pick Alex up.

1.2 Overview

Several studies in Social Psychology suggest that context in human-human social interactions is indeed shaped by culture and gender [14]. These studies suggest that human-computer interactions should be endowed with capabilities to adapt, as in the use case described in the previous section.

Adaptive human-computer interaction is indeed a broad topic in Human-Computer Interaction (HCI) research. Using a software engineering approach, Bouzit et al. proposed the PDA-LPA design space for user interface adaptation [3]. According to these authors, adaptation falls into two categories depending on who is involved in the adaptation process: *adaptability* refers to the user's ability to adapt the interface, whereas *adaptivity* refers to the system's ability to perform interface adaptation. Mixed-initiative adaptation occurs when both the user and the system collaborate in the adaptation process.

Research in intelligent multimodal presentation of information has also tackled interfaces that adapt. Several communication modalities need to be combined dynamically to produce the most relevant output combinations displayed and adapted to the user [39]. This problem involves different concepts related to information structure, interaction components (modes, modalities, devices) and context; it requires the study of the influence of the interaction context on the system's outputs. These authors propose a conceptual model for intelligent multimodal presentation of information that is very relevant for our use case. This model, called WWHT, is based on four concepts: **What, Which, How and Then.** In the following sections we will review these concepts in relation to our use case.

1.3 What to Adapt

Many criteria need to be considered when designing the behaviour of a system to support elderly people in their daily lives. The system should be sensitive to their personal and cultural background. This is quite a challenging task, especially since stereo-typing should be avoided. The use case outlined in Sect. 1.1 clearly shows that two elderly people can be very different in what they would expect from a companion system. Similarly, the notion of culture has to taken with care. While it is common knowledge that, for example, Japanese people are very polite, or English people like black humour, it might be dangerous for a social companion to assume that these traits are true for every single user of that particular country.

Adaptation is an interactive process that needs to take into account the specific user and his or her environment, as well as the system's capabilities for interaction. In the following subsections, we therefore take a closer look at the contextual factors that the system should adapt to and which features of system behaviour can be adapted.

1.3.1 Features of the Context (What to Adapt to)

There are several cases of adaptation described in the use cases where the agent Rob takes decisions based on long-term characteristics of the individuals involved, for example,

- Calling a taxi to take Daniel to his doctor's appointment because it would be too far to walk (health condition);
- Sending an early reminder to Ruth about her appointment as she normally takes longer to get ready (knowledge about a personal characteristic).

Other adaptations are based on transient factors, for example,

- Rob suggests calling a taxi to take Ruth to her doctor's appointment as a heavy storm is predicted (weather conditions);
- Deciding whether Daniel or Ruth can take care of Alex if Eva has a meeting (co-ordination of the calendars of several individuals);
- Sending a text message rather than phoning when the person to be contacted is in a meeting and should not be disturbed (selection of mode of communication based on current availability);
- Contacting by phone as Ruth is currently in the garden and out-of-reach from the smart speakers in the house (selection of mode of communication based on current location).

There is also a more complex scenario in example 3 where Rob infers that Eva cannot pick up Alex from school as usual because she is going to be involved in a meeting. This leads to consultation of the calendars of Daniel and Ruth to find out if either of them can pick up Alex, followed by communicating with Ruth to check that she can do it and then informing Eva that Ruth will pick up Alex. These are just some examples of the sorts of information an agent has to consider in order to be able to adapt intelligently to the needs of individuals.

Looking more generally at adaptivity, and considering in particular the needs of elderly people and multicultural societies, the following are the main features to which an intelligent agent should be able to adapt. Currently, there exist a variety of research projects that address the development of technology for elderly people, but only a few of them address a combination of the features described, for example, Caresses,[1] KRISTINA,[2] EMPATHIC,[3] COADAPT,[4] and ForGenderCare.[5]

[1] http://caressesrobot.org/.

[2] http://kristina.taln.upf.edu/en/.

[3] http://www.empathic-project.eu/.

[4] https://coadapt-project.eu/.

[5] https://www.informatik.uni-augsburg.de/en/chairs/hcm/projects/external/ForGenderCare/.

1.3.1.1 Long-Term Features

Age

Age is potentially an important factor to consider for adaptivity. Generally elderly individuals may show a decline in terms of physical capabilities: for example, in terms of mobility, hearing, sight, concentration, and endurance. However, there is wide variation across individuals depending on factors such as state of health. Age may also have a bearing on other relevant factors such as willingness to adopt new technologies, although again this can vary widely across individuals.

State of Health

The state of an individual's health will determine whether they can perform certain activities, such as cycling or walking long distances. Particular disabilities related to sight, hearing, memory, and ability to concentrate are important factors that need to be considered in terms of adaptivity.

Capabilities and Experience

Some of an individual's capabilities will be determined by the state of their health. There are other capabilities that have been acquired over time, such as familiarity with new technologies, the ability to pick up new ideas quickly, and being able to take in complex instructions and information.

Preferences

Preferences cover a wide range of items, including mode of communication (text messaging, phone, voice with smart speaker, etc.), mode of travel (taxi, bus, car, bicycle, etc.), interests and hobbies, and many other individual preferences.

Personality

An individual's personality influences how they might behave in certain situations. Personality is viewed as a consistent pattern of thought, feelings and actions. OCEAN [6], a well-known model of personality, describes differences in personality in terms of the following traits: openness, conscientiousness, extroversion, agreeableness, and neuroticism. Often only the two main axes of introversion/extraversion and neuroticism/stability are employed.

Culture

Culture is an important factor for adaptivity. There are certain practices and norms that apply in particular cultures and that affect interaction. For example, in some cultures elderly people are shown respect and younger relatives are obliged to defer to the wishes of their elders. Cultural norms also affect terms of address and how individuals engage in interactions in public.

Gender

Linked to culture, gender can play a role in the interaction. For example, there may be differences between males and females in terms of what is permitted in public interactions.

1.3.1.2 Short-Term/Transient Features

Location

Location can affect interaction, for example, whether the interaction is taking place in a public or private location. Location can also affect whether certain modes of communication are available, for example, if WIFI is available in particular locations.

Current Situation/Activity

The current situation describes what the individual is doing at a particular time, for example, taking part in a meeting, busy cooking, etc. This will in turn determine the preferred mode of communication.

Time of Day/Calendar

The time of day can help to determine where an individual might be and what they might be doing. An individual's calendar indicates their availability for various activities.

Emotional State

The individual's current emotional state can affect how the agent communicates with them, for example, if they are depressed, anxious, or happy. In some approaches emotions are divided into a single dimension of positive or negative. Sometimes a second dimension of high and low arousal may also be used.

1.3.2 Features of the Companion System

Consider Example 1 from our use case (reminding about an appointment with the doctor). After taking the decision that a reminder should be given, our hypothetical system needs to further decide how the message should be delivered. In this case the contextual factors described above and in particular the special needs of elderly users from different cultural backgrounds should be taken into account.

Verbal Aspects:

In the domain of elderly care, politeness can be an important factor to avoid potential threats to face, as the feeling of embarrassment or the loss of control over their lives can become important issues for an ageing society. In addition to politeness, the system needs to maintain a certain persuasiveness in order to promote its recommendations. Hammer et al. [20] investigated different verbalizations of the recommendations of a robot assistant to the elderly in terms of its perceived politeness as well as its persuasiveness. To examine potential differences between the target age group and young adults, they tested their system in a laboratory study and within a retirement home. Their results showed that different politeness strategies were indeed perceived differently. The authors therefore highlight that it is important that the verbalization of one and the same recommendation should be chosen with care. In critical situations, for example, the system could emphasize the need to apply a specific action by utilizing *direct commands* that were assessed as being rather impolite, although persuasive.

Cultural background should be taken into account even when non-critical dialogues such as casual small talk conversations are engaged in by a companion system, e.g. to strengthen the social relationship with the user, or to decrease the feeling of loneliness. In [40] typical topics and structures of small talk conversations are outlined. In an analysis of video recordings of German and Japanese small talk conversations, Endrass et al. [13] found significant differences in topic selection and flow of conversations, such as more private topics in German conversations, e.g. hobbies, and more situational topics in Japanese conversations, e.g. the cookies on the table during a coffee break. [11] integrated these differences into culture-specific virtual agent conversations and showed them to human observers from German and Japanese cultures, finding that virtual agent conversations with culture-specific patterns were preferred that were similar to those of their own cultural background.

Para-verbal Aspects:

In cases where the system produces spoken output communication management behaviours are required. Communication management behaviours are used to control the flow of a conversation (e.g. through verbal feedback signals such as "uh-huh" or mutual gaze). These behaviours are culture-specific and can even cause intercultural distress and misunderstandings when used inappropriately [44]. Based on theoretical knowledge as well as empirical findings, Endrass et al. [12] integrated culture-specific usage of pauses in speech as well as overlapping speech into the conversations of virtual agents. The authors found strong trends suggesting that German participants favoured the German versions of the dialogues. In line with the literature, their results suggest that more pauses or more overlaps, compared to prototypical German patterns, were perceived as being disturbing for the German participants. Similar findings were reported by [10], who integrated culture-specific usage of pauses in speech and feedback behaviour for the American and Arabic cultures in conversations with virtual agents. Although their observations are preliminary, culture-related differences were noted in the perception of different usages of silence in speech.

Choice of Device:

Depending on the type of companion system to be developed, e.g. a system that displays messages on a Smart Phone, just generating a speech act might be sufficient. Each device provides certain advantages and disadvantages that should be considered when designing a companion system. While a social robot can enhance messages and suggestions with non-verbal behaviours, supply support on the social level, or apply additional functionality through its physical presence in the user's environment, it comes with quite a high monetary price and is usually not already available in the household where it should provide support. While a Smart Phone is typically available, it has other features that might be difficult in the domain of elderly care (e.g. small displays that are difficult to read with reduced eye sight, or touch interaction that is challenging with shaky hands). Hammer et al. compared seniors' reactions to a health- and well-being promoting recommender system that utilizes a tablet PC or a robotic elderly assistant [19]. Their results suggest that elderly users perceived the robot as more usable than the tablet. The robot was rated, amongst others, as less complex and easier to learn. In terms of the system's persuasion ability no significant difference was found. According to these findings, choosing a social robot seems a good design criterion for a companion system for the elderly. However, the study was conducted within a care home for the elderly in Germany. Whether the same preferences would hold true in other cultures is questionable.

Non-verbal Aspects:

If the companion system should be embodied, e.g. by a virtual agent that is displayed on the companion device or a social robot, appropriate non-verbal behaviours need to be selected to accompany the chosen speech act. This introduces a new design challenge, since non-verbal behaviour can differ vastly across cultures [44]. On the one hand the selection of concrete behaviour can differ (e.g. choosing a friendly hand wave or a bow for a greeting), or the performance of a non-verbal act (e.g. using large, expressive gestures or modest gesturing). Endrass et al. [9] investigated prototypical culture-specific body postures and gestural expressivity for virtual agents in a cross-cultural study in Germany and Japan, finding that some of the investigated aspects, e.g. body posture and speed of gestures indeed have an impact on the users' preferences. In line with the similarity-attraction principle [4], users preferred virtual agent behaviour that resembled their own cultural background. Thus, to enhance understanding and acceptance, the companion system should take the cultural background of the user into account when selecting its non-verbal behaviour.

Simulated Persona:

In order to promote the acceptance by elderly users of a robotic companion that serves as a reminder for appointments, [2] present two different personas (companion versus assistant) for their robotic platform by manipulating verbal and non-verbal behaviour. A study was conducted in assisted living accommodations in which the robot provided appointment reminders. Their results indicate that the companion version of the robot was better accepted and perceived as more likeable and intelligent compared to the assistant version. However, in a long-term study with a robotic companion for elderly users that adapts its spoken language based on explicit human feedback, Ritschel et al. [37] found that an assistant persona was preferred over a companion persona. The contradiction of these findings seems to indicate that personal preferences overrule a general preference for such a companion system by elderly users. This might be an advantage for a robotic system, as it could flexibly switch simulated roles based on the current user's preferences.

1.4 When to Adapt

We now come to the question of when to adapt, which is one of the central questions when designing an adaptive system. We can broadly divide the question into two sub-categories: adaptation *before* the interaction and adaptation *within* an ongoing interaction. If the adaptation takes place before the interaction starts, it is possible to adapt to something that has been set in advance, e.g. in a user profile or in the weather forecast, or to some input from a previous interaction. The latter is of course only possible for frequent interactions. In this way, the system can adapt to something that is static, i.e. the long-term features described in Sect. 1.3.1.1 such as the user's gender or preferences. Consider Example 1 in Sect. 1.1 where the intelligent reminder agent Rob knows that Ruth prefers cycling, whereas Daniel chooses to walk short distances or to take a taxi for longer distances. In contrast, if the adaptation occurs during an ongoing interaction, it is possible to adapt to something that changes over time, i.e. the short-term features described in Sect. 1.3.1.2 like the user's emotion or current situation. Consider Example 2 in Sect. 1.1 where the intelligent reminder agent Rob knows that Eva is at work and therefore should not be interrupted by phone calls. Obviously, this needs to be handled during the ongoing interaction.

Furthermore, when trying to answer the question of when to adapt, the system's decision about when to talk or when to initiate an interaction poses an important subaspect. The system can either act reactively after the user starts the interaction by asking a question or making a request, or it can act proactively without any previous question or inquiry by the user. In this case, the interaction is started by the system, as in Example 1 in Sect. 1.1 (reminding about a doctor's appointment).

Nothdurft et al. [35] describe and analyse the challenges of proactiveness in dialogue systems and how these challenges influence the effectiveness of turn-taking behaviour in multimodal as well as in unimodal dialogue systems. They define proactivity as an autonomous, anticipatory system-initiated behaviour, with the purpose of acting in advance of a future situation, rather than only reacting to it. They focus on proactive behaviour in a mixed-initiative system that combines planning with dialogue by varying the planning between four different variants: a fully autonomous process; adding notifications to the user about the system's decisions; asking the user to confirm decisions; or leaving the decision completely to the user. The authors discuss the challenges in the three different decision making layers *if*, *how*, and *when* and describe a decision space that is constructed by the dimensions *Importance*, *Accuracy*, and *Situation*, which are described as the most important ones for deciding *if* proactive behaviour is necessary. In terms of *how* to intervene, it is suggested that explanations should be provided to foster the building of coherent mental and actual system models. *When* to initiate proactive behaviour seems to be mostly related to the classic turn-taking problem.

Isbell et al. [22] introduce an interface-proactivity (IP) continuum that expresses potential balances of proactivity between the user and the system. At one end of the continuum the user is solely responsible for performing all the actions required to accomplish a particular task. At the other end of the continuum, the system is solely responsible for acting. The authors present adaptive systems and discuss how their characteristics are distributed along the IP continuum. They conclude that the presented continuum allows the description of the design space for adaptive interfaces and helps to identify new parts of the space to explore. Myers and Yorke-Smith [34] characterize the properties desired for the proactive behaviour of intelligent personal assistive agents that can aid a human in performing complex tasks. They define nine principles to guide desired proactive agent behaviour: valuable, pertinent, competent, unobtrusive, transparent, controllable, deferential, anticipatory, and safe. Moreover, an agent cognition model designed to support proactive assistance is presented which employs a meta-level layer to identify potentially helpful actions and determine when it is appropriate to perform them.

A number of proactive systems have already been presented: Yoshino and Kawahara [45] implemented and evaluated a spoken dialogue system that navigates news information and is thereby able to proactively present information related to the user's interest by tracking the user's focus. The authors show that proactive presentations encourage interaction with the system. The virtual selling agent presented by Delecroix et al. [7] is a proactive dialogue agent which initiates the dialogue, uses marketing strategies, and drives the inquiring process for collecting information in order to make relevant proposals. Liao et al. [30] conducted a 17-day field study with a prototype of a personal agent that helps employees find work-related information. The effect of proactive interactions has been explored and the findings show that they carry the risk of interruption, especially for users who are generally averse to interruptions at work, and that an aversion to proactive interactions can be inferred from behavioural signals.

McFarlane and Latorella [32] identify why interruption is an important problem in Human-Computer Interaction and indicate the breadth of applications to which the general problem of interruption management applies. The consequences of poor handling of interruption can have catastrophic results. Therefore, they propose two theoretical frameworks that form a foundation to guide the improvement of interface design for interruption management. Moreover, Jenkins et al. [25] examine the effect of dual-task inference on the interrupting task in response to system-generated interruptive messages. Their findings suggest that the alerts should be bounded in their presentation as the timing of interruptions strongly influences the occurrence of dual-task inference in the brain.

1.5 How to Adapt

1.5.1 Adaptivity in Dialogue

As shown in the scenario in Sect. 1.1, adaptive systems must be able to detect user preferences and needs and use this information to provide personalized services. With this aim, many advances have been made in the multidisciplinary intersection of behavioural research, cognitive science, and computer science in order to accurately predict users' preferences and needs. These advances have been usually applied to marketing (to predict and influence consumer decision making), context-aware recommender systems (e.g. restaurant, travel, tourist guides, music, multimedia, and even job recommenders, etc.), personal assistants and personalized content delivery.

Within these systems, the features described in Sect. 1.3 can be represented as a tuple in which each element takes different values over time. There are also more complex ways of representing this information, for example, using a multidimensional data approach [1]. In both cases the information stored may be fully observable, partially observable or even unobservable. When the information stored is fully observable, it is automatically retrieved by the system and there is a high confidence that the values obtained are correct. For cxample, in the scenario Rob is certain about the user's location because it can be obtained from the smartphone. Sometimes observable information may also be noisy. For example, in the scenario Rob is sure about the appointments reflected in the agenda, but it may be the case that users actually forget to include some of them in their agenda and they would prefer to be asked where they will be at important times (e.g. when they have to pick Daniel from school). The system must then be able to discern which pieces of information may be considered fully or only partially observable.

When the information is partially or not observable, the system must infer the corresponding values, either asking the user (e.g. when Rob is not sure whether the user wants to hear only urgent or all appointments, it explicitly asks the user) or computing them from the context of the interaction (e.g. in Example 3 the system must infer an implicit appointment).

Once the system has observed or inferred all the relevant pieces of information, it must consider them to take appropriate action, including how to proceed with the interaction by generating responses in natural language (either spoken or text). For instance, let's consider the following excerpt from Example 2:

(S1) Rob: Hi Daniel, Eva is asking for babysitting.

(U1) Daniel: When?

(S2) Rob: Next Tuesday from 13:00 to 17:00. You are giving math lessons earlier from 10:00 to 12:00.

(U2) Daniel: Ok, schedule babysitting.

(S3) Rob: Ok, I have scheduled babysitting from 13:00 to 17:00 on Tuesday 3rd November.

As can be observed, after Daniel's utterance (U1), the system must decide how to continue. In this case, the system decides:

- That it must check Eva's agenda to find the time required for babysitting.
- That it should provide the time retrieved and also include information about Daniel's agenda.
- That it should phrase it as shown in S2.

Such decisions could have changed with other users. For example, it could have been less explicit about the user's own agenda.

In conversational systems, this phase is known as *dialogue management*. Dialogue management can be divided into two tasks: (i) dialogue state tracking, that is, deciding what should be the next dialogue state given the current information about the user input and the dialogue history; and (ii) dialogue response selection, finding the appropriate action to be performed based on the selected dialogue state. However, some approaches consider dialogue management as a single monolithic task, and others even merge it with other tasks related to natural language understanding and generation.

Dialogue management can be designed using a rule-based approach in which experts handcraft conversation, or a data-driven approach in which it is learnt from human-human or simulated conversations.

Rule-based approaches allow a fine-grained and fully controlled design in which the system's output is deterministic. Traditionally this has been the most widely accepted option in commercial systems. In a recent book about Conversational UX Design [33], different conversation modules are described that can be reused for different systems, for example, conversation opening, offer of help, repair, conversation closing, and abort, etc. These are reusable patterns that support a more rapid implementation of multiple conversational interfaces. For each of these components, different heuristics and best practices have been detected. For example, what are the best strategies to open the conversation (e.g. name request, offer of help, problem request, capability giving, ...). The drawback of rule-based approaches is that they can be difficult to design in complex tasks and it is complicated to extend them or make them adaptive to new situations. [24] describes in detail these and other aspects to be considered, e.g. the curse of expertise.

Data-driven approaches can be either supervised or unsupervised. User-adaptiveness is a big challenge for these systems, as the corpus used for training the system must contain a sufficiently large number of adaptive behaviours responding to the different features to which the system is adaptive.

Nevertheless, there are different approaches to performing user adaptation in systems that have been trained and optimized for a certain set of users observed within the training corpora. In recommender systems this is usually done by computing an affinity function: users are represented in a space in which each dimension corresponds to an important feature in the application domain (e.g. for restaurant recommendation, the type of food, the price range, ...). When new users arrive they may be either assigned to a random location or asked about their preferences explicitly (do you like Italian food?) or implicitly (would you consider *Il Gondoliere* for a dinner out?) [5].

In data-driven systems user adaptation may involve transferring the knowledge acquired during training in order to make the system responsive to the new user. In this way, instead of having a system that is good for all users on average, it is possible for the system to learn a personalized policy adapted to each user. This task is known as *transfer learning* [27]. For example, [16] defines this task as "transferring in an efficient way relevant information from well-known users called sources to a new user called target". In order to address this challenge, it is necessary to firstly identify the relevant information that can be reused for the new user (source selection), and secondly to include supplementary information and merge it (transition selection).

Following this procedure, [26] presents a negotiation dialogue game in which the preferences of several users must be taken into account over a shared set of options. This is illustrated in the complex setting shown in Example 2 in Sect. 1.1 in which Rob must consider the agendas of Daniel and Ruth in order to respond to Eva's requirements. In this case, user profiles must be enhanced with different parameters that include, for example, the reward for reaching an agreement, a cost distribution, and a cooperation parameter. In our scenario, all users are cooperative (empathic towards Eva's situation) and are inclined to make all efforts to find an agreement, but it may not be the case in all application domains.

Transfer learning is also used to describe the situation in which a system has been trained over a corpus in a domain that is different from the actual application domain in which the system will be used. This is particularly interesting in domains where it is difficult to compile large datasets for training the initial model. [41] presents a recent survey of available corpora for building data-driven dialogue systems.

In conversational systems, the dialogue manager not only has to consider the features described in Sect. 1.3, but also the semantic and dialogic context from previous user and system utterances. There are multiple approaches that involve considering the dialogue history to decide the next system action. [17] introduces context by means of a data structure named the Dialog Register that accounts for the information that has been provided in previous turns. [21] presents context as rules inserted into a knowledge base, in their case the topic of conversation is also a main item of contextual information. In reinforcement learning approaches, dialogue history is encoded into embeddings that are aggregated to obtain a context vector, as described in [18].

In the case of personal assistants that have to be used over sustained periods of time, there are other interesting variables to be considered, for example, disclosure, trust or intimacy, which have an effect on user satisfaction and ultimately in the intention to use the system [28]. In contexts when social information may be crucial, often restrictions must be applied so that the learning of the dialogue strategy is at least semi-supervised. For example, [38] includes pre- and post- conditions to model a behaviour network of conversational strategies that foster rapport. Some authors even distinguish between task strategies and conversational strategies, and the reward for the dialogue manager is not only positive when the task has been successfully completed, but also when the system has been socially effective [23].

Once the next system answer has been decided, it can be translated into a natural language utterance in different ways depending on the context of the interaction. This step, known as *natural language generation* (NLG), can be performed in different ways.

NLG can be divided into different tasks that include: deciding the information to include in the text (content determination); the order in which it is going to be presented (text structuring); how to organize it in sentences (sentence aggregation); finding the right words to use in such sentences (lexicalization); and making the sentences more understandable with the inclusion of references and the construction of well-formed sentences (linguistic realization) [36].

In simple domains this pipeline can be implemented using templates, which is the most widely adopted solution in commercial systems. Templates consist of predefined phrases with slots that are filled with different values depending on the current output. For example, in our scenario, Rob could have predefined phrases such as: "Ok, I have scheduled <TASK> from <HOUR> to <HOUR> on <DATE>". The downside of this approach is its lack of flexibility, as it always produces the same type of output and it is necessary to write new templates for every new system output.

More flexible approaches require some sort of grammar that can be either handcrafted or acquired using data-driven approaches [8]. The latter approach usually understands NLG as an integrated task that is no longer divided into the steps of the pipeline described earlier. This is the dominant approach nowadays. [15] presents an updated survey on NLG that describes the basics as well as recent trends.

NLG can also be learnt in conjunction with the dialogue management task, where the dialogue strategy learnt includes the dialogue act (next system response type) and also the realization of the text [29].

Despite the efforts described, it is generally agreed that personalized spoken dialogue systems have not received much attention and there is significant potential to learn user models along with dialogue models [41].

1.5.2 *Adaptivity in Multimodal Interaction*

Dynamic selection of combinations of modalities such as speech and graphics represent a key entry point for adaptive interaction. Several frameworks have proposed representations for such multimodal interactions. For example, the TYCOON framework [31] defines the following *types of cooperation* between modalities: equivalence, transfer, complementarity, redundancy, and specialization. This typology is quite relevant for the design of adaptive interactions. For example, a given modality might be replaced by another modality which is *equivalent*. Such a *cooperation* between these two modalities enables the system to adapt to different contexts. In another framework [39], mechanisms exploit a behavioural model formalized with adaptation rules based on design goals.

Several multimodal systems have applied such mechanisms for multimodal output. For example, the AdaptO model [43] provides output modalities with the capacity to make decisions, thus collaborating with the fission output mechanism towards a modular solution. It features mechanisms for an adaptable multimodal output system to adapt itself to changing environment conditions (light, noise, distance, etc.) and to its users' needs, limitations and personal choices. Healthcare systems have also applied adaptation. For example, a multi-agent system for tracking and monitoring health data for patients has been developed [42]. These agents use reinforcement learning techniques to build an adaptive user interface for each human user. The actions and behaviour of users are thus monitored and used to modify their respective user interface over time.

1.6 Conclusions

In this chapter, we explored the adaptive nature of systems designed for multicultural and ageing societies. We explained why this adaptivity is required and how it should be tackled along related questions: which system features to adapt, what to adapt to, when to adapt, and how to adapt. We provided an illustrative scenario of a reminder agent in which we focused on these questions by examining a range of adaptive features with a special emphasis on multimodal interaction and dialogue.

Adaptivity needs to be considered along with other main features and design methods for social interactions based on age and culture, including trust, machine learning, and adequate protocols for experimental studies.

References

1. Adomavicius, G., Tuzhilin, A.: Context-aware recommender systems. In: Ricci, F., Rokach, L., Shapira, B., Kantor, P.B. (eds.) Recommender Systems Handbook, pp. 217–253. Springer, US, Boston (2011)

2. Bartl, A., Bosch, S., Brandt, M., Dittrich, M., Lugrin, B.: The influence of a social robot's persona on how it is perceived and accepted by elderly users. In: Agah, A., Cabibihan, J.J., Howard, A.M., Salichs, M.A., He, H. (eds.) 8th International Conference on Social Robotics (ICSR 2016), LNCS, vol. 9979, pp. 681–691. Springer (2016). http://dblp.uni-trier.de/db/conf/socrob/icsr2016.html#BartlBBDL16
3. Bouzit, S., Calvary, G., Coutaz, J., Chêne, D., Petit, E., Vanderdonckt, J.: The pda-lpa design space for user interface adaptation, pp. 353–364 (2017). https://doi.org/10.1109/RCIS.2017.7956559
4. Byrne, D.E.: The Attraction Paradigm. Academic, Cambridge (1971)
5. Christakopoulou, K., Radlinski, F., Hofmann, K.: Towards conversational recommender systems. In: Proceedings of the 22Nd ACM SIGKDD International Conference on Knowledge Discovery and Data Mining, KDD '16, pp. 815–824. ACM, New York, NY, USA (2016)
6. Costa, P.T., McCrae, R.R.: Normal personality assessment in clinical practice: the neo personality inventory. Psychol. Assess. **4**(1), 5 (1992)
7. Delecroix, F., Morge, M., Routier, J.C.: A virtual selling agent which is proactive and adaptive. Advances on Practical Applications of Agents and Multi-Agent Systems, pp. 57–66. Springer, Berlin (2012)
8. Dušek, O., Jurčíček, F.: A context-aware natural language generator for dialogue systems. In: Proceedings of the 17th Annual Meeting of the Special Interest Group on Discourse and Dialogue, pp. 185–190 (2016)
9. Endrass, B., André, E., Rehm, M., Lipi, A.A., Nakano, Y.I.: Culture-related differences in aspects of behavior for virtual characters across Germany and Japan. In: Sonenberg, L., Stone, P., Tumer, K., Yolum, P. (eds.) 10th International Conference on Autonomous Agents and Multiagent Systems (AAMAS 2011), pp. 441–448. IFAAMAS, Richland, SC (2011). http://dblp.uni-trier.de/db/conf/atal/aamas2011.html#EndrassARLN11
10. Endrass, B., Huang, L., André, E., Gratch, J.: A data-driven approach to model Culture-specific communication management styles for virtual agents. In: van der Hoek, W., Kaminka, G.A., Lespérance, Y., Luck, M., Sen, S. (eds.) Proceedings of the 9th International Conference on Autonomous Agents and Multiagent Systems (AAMAS 2010), pp. 99–108 (2010)
11. Endrass, B., Nakano, Y.I., Lipi, A.A., Rehm, M., André, E.: Culture-related topic selection in small talk conversations across Germany and Japan. In: Vilhjálmsson, H.H., Kopp, S., Marsella, S., Thórisson, K.R. (eds.) 11th International Conference on Intelligent Virtual Agents (IVA 2011), LNCS, vol. 6895, pp. 1–13. Springer, Berlin, Heidelberg (2011). http://dblp.uni-trier.de/db/conf/iva/iva2011.html#EndrassNLRA11
12. Endrass, B., Rehm, M., André, E.: Culture-specific communication management for virtual agents. In: Sierra, C., Castelfranchi, C., Decker, K.S., Sichman, J.S. (eds.) 8th International Conference on Autonomous Agents and Multiagent Systems (AAMAS 2009), AAMAS '09, vol. 1, pp. 281–287. IFAAMAS, Richland, SC (2009). http://dblp.uni-trier.de/db/conf/atal/aamas2009-1.html#EndrassRA09
13. Endrass, B., Rehm, M., André, E.: Planning small talk behavior with cultural influences for multiagent systems. Comput. Speech Lang. **25**(2), 158–174 (2011). http://dblp.uni-trier.de/db/journals/csl/csl25.html#EndrassRA11
14. Fischer, A., Rodriguez Mosquera, P., van Vianen, A., Manstead, A.: Gender and culture differences in emotion. Emotion (Washington, D.C.) **4**, 87–94 (2004). https://doi.org/10.1037/1528-3542.4.1.87
15. Gatt, A., Krahmer, E.: Survey of the state of the art in natural language generation: core tasks, applications and evaluation. J. Artif. Intell. Res. **61**, 65–170 (2018)
16. Genevay, A., Laroche, R.: Transfer learning for user adaptation in spoken dialogue systems. In: Proceedings of the 2016 International Conference on Autonomous Agents & Multiagent Systems, AAMAS '16, pp. 975–983. International Foundation for Autonomous Agents and Multiagent Systems, Richland, SC (2016)
17. Griol, D., Callejas, Z.: A Neural Network Approach to Intention Modeling for User-Adapted Conversational Agents (2016)

18. Gupta, R., Rastogi, A., Hakkani-Tur, D.: An Efficient Approach to Encoding Context for Spoken Language Understanding (2018). arXiv:1807.00267 [cs]
19. Hammer, S., Kirchner, K., André, E., Lugrin, B.: Touch or talk?: comparing social robots and tablet Pcs for an elderly assistant recommender system. In: Proceedings of the Companion of the 2017 ACM/IEEE International Conference on Human-Robot Interaction, pp. 129–130. ACM (2017). (nominated for the best poster award)
20. Hammer, S., Lugrin, B., Bogomolov, S., Janowski, K., André, E.: Investigating politeness strategies and their persuasiveness for a robotic elderly assistant. In: Meschtscherjakov, A., de Ruyter, B.E.R., Fuchsberger, V., Murer, M., Tscheligi, M. (eds.) 11th International Conference on Persuasive Technology (PERSUASIVE 2016), LNCS, vol. 9638, pp. 315–326. Springer (2016). http://dblp.uni-trier.de/db/conf/persuasive/persuasive2016.html#HammerLBJA16
21. Herbert, D., Kang, B.H.: Intelligent conversation system using multiple classification ripple down rules and conversational context. Expert Syst. Appl. **112**, 342–352 (2018)
22. Isbell, C.L., Pierce, J.S.: An ip continuum for adaptive interface design. In: Proceedings of HCI International (2005)
23. Jain, A., Pecune, F., Matsuyama, Y., Cassell, J.: A user simulator architecture for socially-aware conversational agents. In: IVA (2018)
24. Janarthanam, S., Lemon, O.: Adaptive generation in dialogue systems using dynamic user modeling. Comput. Linguist. **40**(4), 883–920 (2014)
25. Jenkins, J.L., Anderson, B.B., Vance, A., Kirwan, C.B., Eargle, D.: More harm than good? how messages that interrupt can make us vulnerable. Inf. Syst. Res. **27**(4), 880–896 (2016)
26. Laroche, R., Genevay, A.: A Negotiation Dialogue Game (2017)
27. Lazaric, A.: Transfer in Reinforcement Learning: a Framework and a Survey **12**, 143–173 (2012)
28. Lee, S., Choi, J.: Enhancing user experience with conversational agent for movie recommendation: effects of self-disclosure and reciprocity. Int. J. Hum. Comput Stud. **103**, 95–105 (2017)
29. Lemon, O.: Learning what to say and how to say it: joint optimisation of spoken dialogue management and natural language generation. Comput. Speech Lang. **25**(2), 210–221 (2011)
30. Liao, Q.V., Davis, M., Geyer, W., Muller, M., Shami, N.S.: What can you do?: studying social-agent orientation and agent proactive interactions with an agent for employees. In: Proceedings of the 2016 ACM Conference on Designing Interactive Systems, pp. 264–275. ACM (2016)
31. Martin, J.C., Béroule, D.: Types of cooperation for observing, evaluating and specifying cooperations. In: Proceedings of the AAAI Fall 1999 Symposium on Psychological Models of Communication in Collaborative Systems
32. McFarlane, D.C., Latorella, K.A.: The scope and importance of human interruption in human-computer interaction design. Hum.-Comput. Interact. **17**(1), 1–61 (2002)
33. Moore, R.J., Szymanski, M.H., Arar, R., Ren, G.J. (eds.): Studies in Conversational UX Design. Human-Computer Interaction Series. Springer International Publishing, Cham (2018). https://doi.org/10.1007/978-3-319-95579-7
34. Myers, K., Yorke-Smith, N.: Proactive behavior of a personal assistive agent. In: Proceedings of the AAMAS Workshop on Metareasoning in Agent-Based Systems, Honolulu, HI, pp. 31–45 (2007)
35. Nothdurft, F., Ultes, S., Minker, W.: Finding appropriate interaction strategies for proactive dialogue systems - an open quest. In: Proceedings of the 2nd European and the 5th Nordic Symposium on Multimodal Communication, 6–8 Aug 2014, Tartu, Estonia, 110, pp. 73–80 (2015)
36. Reiter, E., Dale, R.: Building Natural Language Generation Systems. Studies in Natural Language Processing. Cambridge University Press, Cambridge (2000). https://doi.org/10.1017/CBO9780511519857
37. Ritschel, H., Seiderer, A., Janowsk, K., Wagner, S., André, E.: Adaptive linguistic style for an assistive robotic health companion based on explicit human feedback. In: 12th PErvasive Technologies Related to Assistive Environments Conference (PETRA '19) (2019)
38. Romero, O.J., Zhao, R., Cassell, J.: Cognitive-Inspired Conversational-Strategy Reasoner for Socially-Aware Agents pp. 3807–3813 (2017)

39. Rousseau, C., Bellik, Y., Vernier, F., Bazalgette, D.: A framework for the intelligent multimodal presentation of information. Signal Process. **86**(12), 3696–3713 (2006). http://dx.doi.org/10.1016/j.sigpro.2006.02.041
40. Schneider, K.P.: Small Talk: Analysing Phatic Discourse. Hitzeroth, Marburg (1988)
41. Serban, I.V., Lowe, R., Henderson, P., Charlin, L., Pineau, J.: A Survey of Available Corpora for Building Data-Driven Dialogue Systems (2015). arXiv:1512.05742 [cs, stat]
42. Shakshuki, E., Reid, M., Sheltami, T.: An adaptive user interface in healthcare. Procedia Comput. Sci. **56**, 49–58 (2015). https://doi.org/10.1016/j.procs.2015.07.182
43. Teixeira, A., Pereira, C., Silva, M., Pacheco, O., Neves, A., Pereira, J.: Adapto - adaptive multimodal output. In: Proceedings of the AAAI Fall 1999 Symposium on Psychological Models of Communication in Collaborative Systems
44. Ting-Toomey, S.: Communicating Across Cultures. The Guilford Press, New York (1999)
45. Yoshino, K., Kawahara, T.: News navigation system based on proactive dialogue strategy. In: Natural Language Dialog Systems and Intelligent Assistants, pp. 15–25. Springer (2015)

Chapter 2
Modeling Trust and Empathy for Socially Interactive Robots

Patrick Gebhard, Ruth Aylett, Ryuichiro Higashinaka, Kristiina Jokinen, Hiroki Tanaka, and Koichiro Yoshino

2.1 Introduction

Current society is rapidly changing. Not only increased digitalization of various services affect our work and everyday life but also various changes in the physical environment, demography and migration of the population, changes in the work situations, transportation, news delivery, etc., have a big impact on the way we work and interact with other humans. In this context, novel AI technology has been greeted as a way to resolve problematic issues in elder-care, education, transportation, etc., situations, and indeed, remarkable progress has been made in applying AI techniques to medical and bioscience problems, autonomous driving, entertainment, computer vision, and language translation problems. The development of robotics has also advanced AI-based solutions, and the development has not only focused on the use

P. Gebhard (✉)
German Research Center for Artificial Intelligence, Kaiserslautern, Germany
e-mail: patrick.gebhard@dfki.de

R. Aylett
Heriot-Watt University, Edinburgh, Scotland
e-mail: r.s.aylett@hw.ac.uk

R. Higashinaka
Nagoya University, Nagoya, Japan
e-mail: higashinaka@i.nagoya-u.ac.jp

K. Jokinen
National Institute of Advanced Industrial Science and Technology, Tokyo, Japan
e-mail: kristiina.jokinen@aist.go.jp

H. Tanaka · K. Yoshino
Nara Institute of Science and Technology, Ikoma, Japan
e-mail: hiroki-tan@is.naist.jp

K. Yoshino
e-mail: koichiro@is.naist.jp

J. Miehle et al. (eds.), *Multimodal Agents for Ageing and Multicultural Societies*,
https://doi.org/10.1007/978-981-16-3476-5_2

of robots as tools in manufacturing industry, but as assistants and social companions to humans in various care-giving and educational settings. Social Robotics has rapidly developed toward a possible solution to the changes in the modern society, in particular to address the issues concerning aging population and deceasing number of care-givers, being integrated with IoT technology in smart home and smart city environments, and forming a new type of social agent as a means of instant communication.

As the need for interactive systems in various service sectors increases along with the digitalization in the society, it is important to develop solutions that take the human user and human interaction patterns into consideration. This opens up development of novel types of interactive systems where the robot is not just a tool but an agent to communicate with (Jokinen 2018). Humanoid robots which can move in the three-dimensional environment and observe the human participants, can embody interaction which shows more sensitivity to social aspects of communication: in the ideal case, social robots can understand people's needs and emotions, communicate in natural language, and offer peer-type assistance and companionship with their capability to chat, monitor, and provide information.

In practical care-work, the goal of robot assistants is much less ambitious, e.g., to assist in patient registration or to instruct novice care-givers how to do a particular task. However, the robot agent's communication ability is an important functionality that influences how the robot is accepted by the users and how reliable it is seen as delivering information that the user has no direct contact with. In particular, applications that collect medical, biosensor, or private information from the user need to be careful in supporting safe and secure data collection, data delivery, and considerate, trustworthy and empathic use of data in the interactive situations with the user. Many of the safety issues can be handled with the standard data security and encryption techniques, but the issues concerning interaction with the user and the potential evesdroppers need to be paid outmost attention, due to sensitivity of data. Ability of such systems to provide competent and empathic interaction is crucial when considering how the human users can develop trust in the system's capabilities and empathic

Intelligent social robot applications that allow the users and carers to talk about the user's personal information, create challenges for the development of multimodal dialogue system on several levels, ranging from to dialogue management and trust in delivering truthful information. On one hand, connectivity and sensor technology is needed to create and improve services for care-takers and elder people so as to assist their well-being in daily life and increase safety in care-taking and monitoring in general; on the other hand, such information collected, used, and stored by the robot needs to be carefully handled to prevent data security breach and to protect personal sensitive information of the users. For instance, only authorized users like doctors and nurses can have access to the individual user's medical history, and the robot must also respect these requirements. This can be handled by access rules that allow only certain people and user groups to check the database for a particular user's medical records. However, at the same time, the social robot needs to take into account the needs for smooth interaction. It should be able to recognize its interaction partner by a combination of face, voice, and motion recognition technology rather than requiring the user to type in a password on a computer screen.

In dialogue situations, an important issue is that the robot should be able to assess the context in which it can deliver information that is sensitive or which the user concerns private. For instance, the robot should not give answers to questions that concern e.g., the neighbor's private information, or what the doctor has talked with the previous patient. Moreover, in a multiparty situation, the robot agent must respect privacy requirements for each partner so as not to disclose critical and sensitive information to those who have no right to it. Consequently, the robot's perception capability of the environment may need to be enhanced with information from the motion tracker and recognition devices installed in the environment where the interaction takes place. Such a scenario of an intelligent home or a care-facility may not be too complicated technologically, but raises further issues on privacy and data protection. Furthermore, the inter-connected devices will need to share information about the persons in the share information about the persons in the intelligent space, so the robot needs to apply similar rules of privacy protection to its interaction with other robots and IoT than is required in its interaction with the human users. In fact, the whole background information processing and exchange must be governed by the same access rules as the robot's interaction with the human users: some types of robot agents and classes of services have no access to a particular type of information while some do. The main question then is who gives the access rights to the devices and who decides which groups of types of services the information can be shared.

2.1.1 Motivating Distinctions

Trust and empathy are typical concepts that seem to be clear and well-understood if we use them in referring to truthful and reliable behavior or emotionally rewarding and sympathetic attitudes, but their clarity soon vanishes if we try to define them with respect to practical examples. The positively valued concepts tend to become blurred due to complexity of the issues involved in their definitions, and the highly subjective views of the behavioral and cognitive aspects that are defining features of appropriate manifestation of the meaning of the two concepts.

In this chapter we study these concepts in the context of social robots and human–agent interaction, where the concepts are being discussed concerning the user's trust in the robot's capability to produce acceptable and trustworthy behavior or furnish the human partner with truthful information.

We distinguish two dimensions in which trust and empathy can be divided: behavioral aspects and cognitive aspects. While these aspects may also be problematic in a detailed sense, they provide, however, a clear enough way to distinguish the two dimensions that are to be defined. The behavioral aspects of trust and empathy refer to the agent's behavior that distinguished trustworthy and emphatic behaviors from those which are not. For instance, trusting that a robot is able to lift a patient without dropping the person or squeezing them too tight is trust of the agent's skills to be able to perform the task correctly and fully. In the field of human–computer interaction design, such requirement is called usability: the system (in this case, a robot agent) is capable of performing the function it is designed for. If the robot agent's behavior is perceived as being related to their benevolent attitudes and decision making (when they also have power and capability to behave in the opposite way), and which then manifests in their supportive and affective behavior, then the agent can be said to exhibit emphatic behavior. On the other hand, if the agent's behavior is related to its cognitive capabilities, its ability to deliver truthful information, provide enjoyable conversation, draw correct conclusions, and in general show capability of intelligent reasoning, then we can talk about cognitive trust, i.e., trust in the agent's cognitive competence. Similarly, if the agent can be said to exhibit cognitive empathy, this means that the agent can provide words and gestures that show understanding and sharing of the same experience, feelings, thoughts that the partner describes.

2.1.2 Applications Domains for Elderly People

2.1.2.1 Dementia Trip Friend

One of the possible applications will be "Dementia Trip Friend" (DTF), which accompanies a user's trip and assists the user having cognitive impairment such as dementia. It is recognized that outing, such as going on a trip, decreases as people become older. One of the possible reasons is that they tend to live alone and do not have company for going out. A DTF can be a good company and it can motivate the user for going out through appropriate recommendations.

The DTF can help users during and after a trip. During the trip, the system can converse with users, performing social interaction, which can decrease the risk of dementia. It can also make the trip more enjoyable through empathic conversations. The system can monitor the user's cognitive function by monitoring the user with various devices. Chen et al. suggests that the cognitive function can be accurately estimated by using a combination of wearable devices [23]. The DTF can also support the user when the user needs assistance and communicate with care-givers when necessary. The user is also reported by the system with his/her cognitive status whenever necessary, which can motive the user for more outing if certain skills need to be improved. After the trip, it can help the user to look back on the trip and their shared memory; realizing the opportunity for the user to have social interaction. Otaki and Ohtake found that it is possible for a user and an automated system to share a story through a co-imagination method, which is typically conducted between humans having cognitive impairment [98]. In the conversation, it may be a good opportunity for the system to monitor the cognitive skills of the user; for example, in addition to the acoustic features of the user's speech [68], reactions to a variety of questions are found to be a good clue to detect cognitive impairment [126].

2.1.2.2 Empathic Family Companion

In a not so distant future, social robots might be part of our daily life at home. Their task is to support us in our activities. For elderly people, this might mean having a care robot that overtakes some tasks that the person cannot do alone anymore. One challenge that might occur when getting older is incontinence and the need for incontinence aids. Many elderly persons cannot assess if it is needed to change the incontinence aids. A robot, equipped with olfactory sensors, could gauge that need and address it to the elderly person, where it is vital that this is done in an empathic way. The robot should know about the context situation, e.g., if the elderly person is alone or has guests, and adapt accordingly. In case the person is not alone, the robot has to know that it might not be appropriate to tell the elderly person in front of his guests that the incontinence aids have to be changed. Therefore, it needs an understanding of what information should stay private between the robot and the elderly person and what can be addressed in others' presence. In the case of other people's presence, a human caretaker for elderly people would get close to the elderly person and try to ask him quietly to join for changing the incontinence aids. The idea behind this is to keep one's privacy and to avoid that the person might feel ashamed. This knowledge should also be given to a care robot. The robot has to be able to understand that in some situations, the elderly person might feel ashamed for various caring aspects, e.g., those connected to privacy. Therefore, the robot needs a theory of mind about the emotions evoked in the elderly person in different situations. In the situation that the incontinence aids have to be changed, the robot has to identify it as a potentially shame eliciting situation and has to find a way to deliver his message discreetly.

Within that domain, other possible scenarios are in which people (1) are losing their eyesight (or different bodily abilities), (2) are losing cognitive abilities, and (3) are losing partners and friends (their peer-group). Since all these three cases are existential parts of our social life, It could be assumed that if they occur, they trigger negative emotions. A social robot that has been known for a long time and is in the role of a companion might be able to give support on an emotional level and compensate disabilities (e.g., reading the news or a book if eyesight diminishes, supporting difficult decisions, or helping to connect to other people). In any case, the robot must know about individual and cultural values and norms to "guess" the appropriate level of support. Situations of paternalism should be avoided. The robot's given support should be explained by the robot adapting to the current state of the human. This includes the human's current emotional state, her or his abilities to regulate this state, and the current social and environmental situation (context). A significant task could be providing external interpersonal emotion regulation to allow the human to establish an emotional balance.

2.1.2.3 Assistive and Supportive Technology

Kaliouby et al. designed the technology to help children on the autism spectrum better understand how other people felt because they struggle with making eye contact or reading people's expression of emotion [42]. Their group developed a system that intuitively presents various emotional interpretations of the others' facial-head movement [81]. Also, physiological responses such as heart rate, skin conductance, and brain wave might help people with autism spectrum learn how to understand and regulate [89] their own physiological activation without having to share their internal changes with others [103].

On the other hand, some papers proposed to support the expression of non-verbal and verbal components. Job interviews have also received a lot of interest from researchers in Affective Computing for training candidates [62]. This could be applied to social skills training in people with autism spectrum [127]. Such virtual training tools are limited to a one-to-one interaction with a virtual recruiter. Several Virtual Reality tools for training public speaking are already available on the market. Also, previous work includes real-time or simultaneous feedback [128].

Such technology has the possibility to be applied to support elderly people with autism spectrum or with cognitive impairments who have difficulties in understanding and expressing emotion. Cross-cultural comparisons are essential for persons with autism spectrum and are discussed in [84] (Greece, Italy, Japan, Poland, and the United States) while cultural differences may impact the perception of autistic symptoms.

If such systems correctly provide recognized emotion and users can understand them, users will satisfy and trust in the system.

2.2 Background

Trust and empathy are two of the key elements in social agents.

Trust is important because it is essential to make communication efficient. If humans and computers do not trust each other, it would lead to an unnecessary cost, arising from protecting oneself from various threats, such as deception, privacy violation, and so forth [70]. As the development of artificial intelligence technologies progress, trust modeling between computer and machines becomes an important issue for better human–computer communication.

Mayer et al. list competence, benevolence, and integrity as the main factors of trust [88]. Their definition of trust is widely used:"willingness of a party to be vulnerable to the actions of another party based on the expectation that the other will perform a particular action important to the truster; irrespective of the ability to monitor or control that party." This definition highlights two important aspects of trust; one is risk (vulnerability) and other is the willingness to take that risk on the basis of the expected behavior of the trustee.

The source of risk can come from the competence of a system. Hancock et al. identified human-related, robot-related, and environmental factors in human–robot interaction [54]. According to their study, robot performance (competence) had the highest impact on trust. Abd et al. also performed an experiment in which a user performed a collaborative task with an assistant robot [1]. They also found that the system competence (e.g., accuracy of operations) was most related to trust.

Such risk can be avoided/taken by having an accurate expectation (predictive model). Expectation can come from various sources, including credentials (certificate of trust give by other parties), reputation (past behavior), personality (one's preferences), and social/cultural norms (how one would behave in a society). If agents can act as expected by users, it becomes easier for human users to believe in the agents' judgments [21]. Especially in the multi-cultural society, culture/social norms must be considered because expected behavior changes from culture to culture [76]. There is work in human–human communication that the level of trust changes depending on the nationality [37]. Machines that interact with humans need to respect the culture in which they operate.

Trust can be categorized into initial trust and continuous trust. Initial trust means the trust that the user has on the system before using the system. In this situation, elements such as the representation and reputation are important [119]. In continuous trust, factors such as the performance, sociability, security, and mutual goals are important. In addition, threats related to job replacement also need to be considered. There is also work that categorizes trust into cognitive trust and affective trust [80]. The former is based on one's cognition of the system and the latter on one's willingness to use the system irrespective of one's understanding of the system, which is related initial trust.

In the context of elderly care, trust becomes more important because the elderly tend to be more vulnerable than younger population. A high level of trust has to be established and maintained between the elderly and the elderly care agent. In addition, we must think about the people who care the elderly, including their families and medical personnel. The trust has to be made between many stakeholders. Cultures and social norms have to be taken care of by the system so that the elderly and the carers can have reasonable expectations about the system.

In [26], a questionnaire-based research was performed to examine what parents and care-givers of autism children think are important for care-giving agents. They found that, "supervised autonomy,'' which means the system's autonomy under the supervision of care-givers [130], is preferred. In addition, many parents and care-givers thought that the looks of the agents should not look like realistic humans. There seems to be a fear that elements of human-to-human communication is lost by the intervention of an agent. Although this study does not involve the elderly, this highlights the need to take into account the opinions of many stakeholders.

With regards to empathy, it is an actively studied subject in human–computer interaction. According to Hoffman, empathy is defined as "a psychological process that makes a person have feelings that are more congruent with another's situation than with his own situation" [58]. Hoffman also introduced the division of cognitive empathy and affective empathy [57]. The former is the ability to understand another's mental state and the latter is the ability to react affectively to another's emotional state. This distinction is widely adopted in human–computer interaction [101]. Empathy is regarded as a key factor in the human society because it enables us to understand others and to increase social bonding. This is also true in human–computer society, as Paiva et al. put it, "a social agent—that is, an agent with social ability—must be endowed with the capability to understand others and their intentions, motivations, and feelings.''

Empathy is obviously one of the factors affecting trust between humans and machines because understanding mental states of users is definitely related to the performance (competence) of social agents. Empathy is also related to expectation because it is about having a model of others. If the system has the capability of acting according to the expectation of a user, it is likely that trust can be cultivated. There has been work that examines the relationship between trust and empathy in human–computer interaction. For example, Cramer et al. performed an experiment using videos showing an interaction between an robot with a human, and found that an empathic robot was perceived more dependable than a non-empathic robot [29]. It was also found that the empathic accuracy (the accuracy of understanding user's emotion) had an effect on the perceived credibility. In [20], in an experiment using Casino-style blackjack game between a user and system, it was found that showing an empathic emotion had a significant effect on the trustworthiness of the agent. These results indicate that empathy is indeed one of the factors affecting trust.

For elderly care, the role of empathy is important because the task is more related to mental than physical. In a multi-cultural society, empathy is also important because it enables the understanding of others, thus leads to better expectation. The empathic capability and the trust that builds upon it would be the key elements in realizing socially interactive robots in the multi-cultural aging society.

2.3 Modeling Trust

The importance of trust in the domains covered by this book can be seen from a recent definition [123] the: *willingness of a party to be vulnerable to the actions of another party based on the expectation that the other will perform a particular action important to the trustor, irrespective of the ability to monitor or control that party*. In elder support, it is frequently the case that the person being supported is already vulnerable, willing or not, because of cognitive decay, physical restrictions, or both.

Not only this, the specific applications often considered as targets involve both physical intimacy—help with dressing, eating, washing and other everyday tasks—and emotional intimacy—companionship, emotional support, and empathic engagement. In both domains, greater intimacy also implies greater vulnerability. This is doubly true in explicitly therapeutic applications, for example post-stroke exercise, or acting as a cognitive prosthesis to an elderly dementia sufferer.

One should not assume that the *trustor* is necessarily only the elderly person themselves. Their vulnerable status may require human guardianship from a professional carer or one or more family members, who would have to exercise trust on their behalf. A system acting as an assistant to a human carer would probably have to be trusted by multiple people. Carers may also have to trust such a system in situations when they may not be present themselves to observe its behavior—for example, to act for them when they are out doing shopping or meeting a friend. Thus the *lack of ability to monitor or control that party* may be due to absence as well as to vulnerability.

The importance of trust is therefore clear, but there are at least three different possible reasons for modeling it. The first and most important is if this helps us to build *trustworthy* systems, that can be relied on to carry out intimate tasks with vulnerable people. Here the key issue is surely the ability to assess how far a system is trustworthy, requiring appropriate metrics. We can think of this as a static attribute of the system. That the system itself needs a model of its own trustworthiness is clear if one considers the problem of *breaking trust*, or as we would term it, betrayal. This is clearly to be avoided.

Two other reasons relate to interaction and a dynamic view of trust. Firstly, could the system assess how far it is trusted by a specific person in a specific context? This is a dynamic recognition task, and like most such recognition tasks, a very difficult one, even conceptually. However without this, a system might inadvertently break trust and never notice it had done so.

Secondly, could a system then adjust the degree of trust it invokes in a specific person in a specific context? Recovery from an error reducing trust might be one reason for doing this. However adjustment need not mean *increasing* the extent to which the system is trusted since to produce trust greater than its actual degree of trustworthiness—*overtrust*is clearly ethically problematic. As we will see below, overtrust is a very real issue.

A final reason for modeling trust might be if we want a system that can display trust in the person it is interacting with. This may in fact be a component of the dynamic adjustment of trust already referred to, since it could be that we trust more those we believe trust us.

Modeling of trust began in a very different setting, investigating relationships within organizations and between managers and those they managed. Mayer et al. [88] carried out seminal work to construct a model that can be seen in Fig. 2.1. Given the organizational context in which this model was formulated, trust here refers to trust with respect to a specific task. *Ability* is competence in carrying out the task, *Benevolence* is stance in relation to the trustor's goals—a commitment to carry them out, and *Integrity*is consistent adherence to social or organizational norms. Trust is seen as influenced both by the degree of risk the trustor is willing to take in a relationship and by the perceived risk to the task if trust is misplaced.

It is worth pointing out that this is a very high-level model, some way from an operationalization for actual systems. A similar point could be made about the subsequent model developed in the context of robot teleoperation systems [54]. This is a step towards the domains of this book in that it relates to human and robot rather than

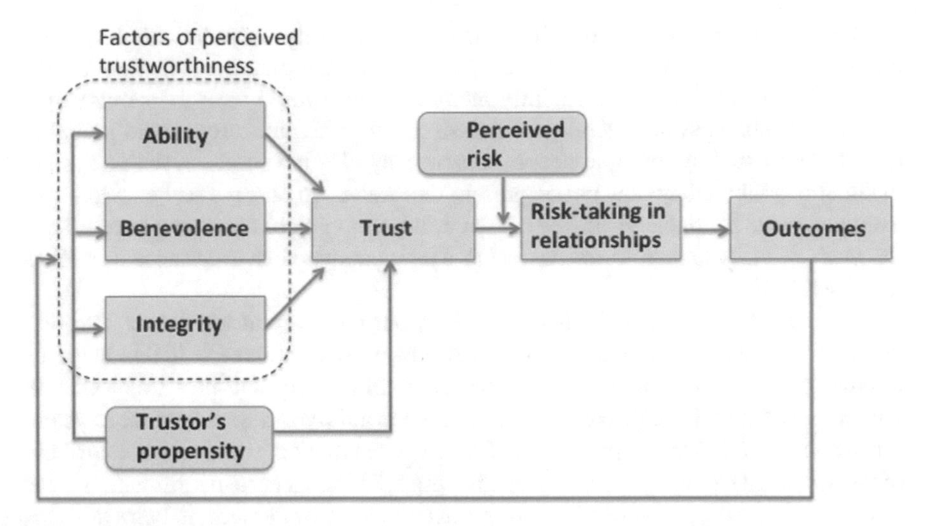

Fig. 2.1 Mayer's factors of trust model, see: [88]

human and human, and was based on a meta-analysis of work in this area. Figure 2.2, taken from [54], shows a very large set of attributes grouped under those relating to the human tele-operator, those related to the robot, and those related to the environment in which tele-operation takes place. These attributes are more concrete than those in the Meyer model but offer nothing on relative importance, inter-relation, or operationalization—for instance what behavior exactly is of interest in the Robot attribute *behavior*. Noticeably, where Mayer has *Benevolence* and *Integrity*, Hancock's attributes are not easily relatable to either, unless through the Robot attribute *Robot Personality*. This is not so surprising when you consider tele-operation involves robots with zero or little autonomy and is focused on the achievement of a specific task.

This model does however include *Transparency* as an important performance-based attribute, and one that might be used in interaction by a human to judge the system's benevolence and integrity. Indeed, the set of performance-based attributes could offer a basis for an interaction-based model of trust that would look at how trust is established, maintained and adjusted in interaction. Subsequent work by this group [113] draws on the literature to associate the tabulated attributes with dynamic trust. Figure 2.3 specifies some of these attributes against changes in the level of trust. We should note from this table that anthropomorphism has been studied more than once and has a noticeable impact on trust.

One should add that *trust* is a term used in more than one way in the literature. In particular, its usage in data communications, multi-agent systems and the Internet of Things differs from the definition of [123]. Thus [72] lists:

- Comfort of Use (*The system should be easy to handle*)
- Transparency (*I need to understand what the system is doing*)

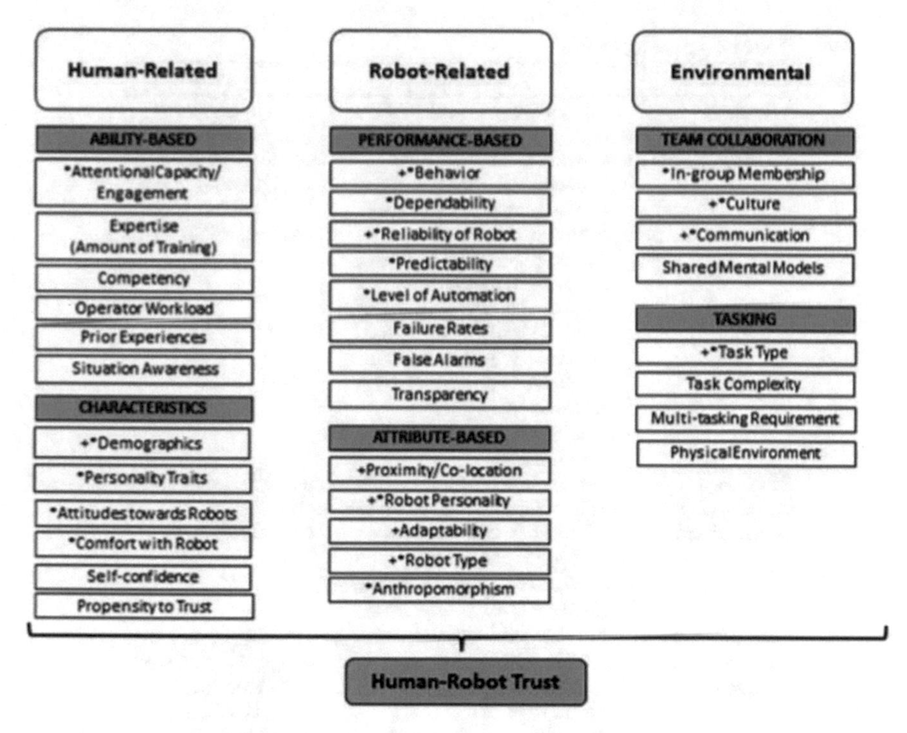

Fig. 2.2 Hancock's triadic model of trust, see: [54]

- Controllability (*I want to be in control of the system's actions*)
- Privacy (*The system should neither ask for nor reveal private information*)
- Reliability (*The system should run in a stable manner*)
- Security (*The system should safely transfer data*)
- Credibility (*The system should have been recommended by others*)
- Seriousness (*The system should have a professional appearance*).

This overlaps very little with [54] and in the view of [71] is closer to *credibility* than to trust per se. A similar difference can be seen in [80], whose questionnaire uses the definition: *the extent to which a user is confident in, and willing to act on the basis of, the recommendations, actions, and decisions of an artificially intelligent decision aid.*

A major difference is that the tradition in which [72] is working deals with *disembodied* agents, in which the humanlike aspects of interaction and the ascription of social agency are much reduced or even non-existent.

Kulms and Kopp [71] investigated the impact of embodiment and competence over time in a cooperative game in which the human player received advice on their move from an agent. Their study showed that initially, an embodied agent invoked

Robot Characteristics

Characteristics	Outcome	Support
Behavior / Robot Type/ Anthropomorphism	• Trust increases when robot behavior, appearance and type matches what is expected • A companion robot should be considerate, proactive, non-intrusive, flexible, and competent, to work towards a relationship of trust and confidentiality with the human • Robots should express social behaviors (e.g., turn taking, emotional expressions) to be trustworthy • Anthropomorphism is correlated with trust	(Dautenhahn, 2007; Evers, Maldanado, Brodecki, & Hinds, 2008; Looije, Neerincx, & Cnossen, 2010)
Reliability/ Predictability/ Dependability	• A robot's lack of self-awareness in dangerous situations leads to a distrust of robot teammates • Increase in reliability, predictability & dependability of robot actions increases trust	(Kidd & Breazeal, 2004; Groom & Nass, 2007; Ross, 2008)
Level of Automation/ Adaptability	• Increased trust leads to increased reliance on automation • To give autonomy to machines, we must be able to trust the underlying processes and mechanisms, as well as the fault-detection mechanisms • Trust increases when human has the ability to adapt level of automation	(Tenney, Rogers, & Pew, 1998; Steinfeld et al, 2006; LeBlanc, 2007; Heerink, Krose, & Evers, 2010)
Proximity	• Increased trust with robots that are co-located	(Bainbridge, Hart, Kim, & Scassellati, 2008)
Robot Personality	• A highly likable robot with active response and engagement correlates with trust	(de Ruyter, Saini, Markopoulou, & van Breemen, 2005; Kidd & Breazeal, 2008; Rau, Li, & Li, 2009; Li, Rau, & Li, 2010)

Fig. 2.3 Theorized and empirically supported influences of trust: robot characteristics, see: [113]

a higher level of trust than a non-embodied agent, but that trust was adjusted in line with competence over time. This supports the idea that *First Impressions* have an important effect on initial trust, but that trust varies dynamically in interaction.

2.3.1 Overtrust

So far we have seen an emphasis on task competence in approaches to trust, which we can relate to the situations for which they were developed. However a series of studies in human–robot trust have shown that trust can remain high even when the robot is far from competent at the task (overtrust), underlining the importance of social factors—much less emphasized in these models—on the regulation and maintenance of trust.

Robinette et al. [110] reports work in which a robot with demonstrably poor navigational skills was trusted to lead people out of a (realistically) simulated fire. Though not specifically investigated, one might hypothesize a level of status or authority was attributed to the robot. Booth et al. [19] ran a study that demonstrated students would allow a robot to enter a dorm behind them even when there were some signaled grounds for suspecting it intended harm.

Ullman and Malle [132] showed that allowing a human to start robot plan execution with a button press raised their level of trust in the robot's ability to run the plan successfully. Hamacher et al. [52] compared three versions of a robot, one that was inexpressive and competent, one that was inexpressive, made a mistake, and attempted to rectify it, and one that was communicative, expressive, and also made a mistake and tried to rectify it. Comparing the inexpressive robots, participants showed more trust in the more efficient one. However the expressive robot was trusted the most in spite of its error, and its ability to make an expressive apology even appeared to affect perception of task time, since it was perceived as the quickest to complete the task even though the error and its extra interaction made it in fact the slowest.

This work suggests that research into *rapport* [144] might be relevant to understanding the development and maintenance of trust. While work on trust is closely connected to task execution, work on rapport studies the social element of interaction, in particular examining the processes of Mutual Attentiveness, *Face* management, and Co-ordination. Some of the factors examined in this model of rapport, such as self-disclosure, reciprocity and praise, could easily be transposed into the models of trust already discussed. Rapport is also closely related to empathy, especially in the area of reciprocity.

Incorporating a stronger account of social interaction processes would also require us to take the impact of different cultures and social norms/values upon trust more seriously. While individual variations already appear as an attribute in the discussed models, group differences do not. Given we focus on trust in systems supporting the elderly, generational variations are also of interest.

We have already mentioned that systems supporting the elderly are likely to be involved with other people around them, be they professional carers or family and friends. This raises the possibility of conflicts in what constitutes trustworthy behavior. An obvious arena for such conflicts is that of privacy and autonomy: an elderly person may trust a companion robot to keep information about their condition and activities private that carers might expect the same robot to disclose. Indeed [134] argues that *forgetting*, in the sense of removing raw data in favor of generalizations over many episodes, is necessary from an ethical perspective.

Arguing that trust is related to benevolence as in [88] does not solve the problem of what in specific situations counts as benevolent. Thus ethical issues are closely linked to trust.

2.4 Modeling Empathy

People are social and in need of relationships. Empathy plays a central role within societies and across cultural borders of societies. Empathy is a complex human concept that manifests itself in the ability to show empathic behavior. Empathy is connected to the human ability to build a theory of mind.

Interactive agents have not necessarily been empathic. It is unclear to what extent physical care robots (e.g., carrying or washing robots) have to be empathic for some tasks. For sure, they have to act carefully. It could be argued that the more such agents are used in social environments in the role of assistants, partners, or even experts, the more such technology should come with models of how human social life is structured and organized. However, up to what extent technology can or should "act" emphatically?

2.4.1 Overview, Challenges, and Definitions

For this work, we assume that empathic cultural-aware agents are (anthropomorphic looking) machines. Such agents can be seen as a tool that is designed to simulate aspects of human cultural-aware empathy.

Since Aristoteles, empathy is often described as being related to how others experience a situation and others' emotions. There is no universal definition of empathy [39]. Empathy could be defined as a set of abilities, such as perspective-taking, transposing oneself imaginatively into fictitious characters' feelings and actions, assessing feelings of others, or uneasiness intense interpersonal settings [32]. Empathy could be defined from a process-orient perspective, such as "the involvement of psychological processes that make a person have feelings that are more congruent with another's situation than with his situation" [58].

One traditional definition of empathy distinguishes between cognitive empathy and affective empathy (e.g., [40, 56, 58], as mentioned in [6]). The cognitive aspect addresses the ability to create or infer other people's thoughts (beliefs, desires, intentions) and involved cognitive processes. Affective empathy usually addresses the individuals' emotional reactions in response to another person's negative feelings. This differentiation of empathy is often used to computationally model empathy for interactive agents (e.g., [101]).

Empathic cultural-aware Agents

Researchers in the area of empathic agents are motivated by several reasons why such agents should be empathic. A general motivation is that agents are more likely to be accepted if they know the user as a social actor [104, p. 247]. This approach includes that agents should act in a social (familiar) way. Moreover, they might be able to show empathic behavior (e.g. [16, 129]. Related are the research questions: (1) if the agent behavior is perceived as social (empathic) behavior and (2) if the believability of such agents is increased by such (e.g. [30, 77, 100, 133]). To do so, such agents must come with the ability to understand others at the level of intentions, motivations, and feelings, which includes perceiving and understanding others' affective states and acting accordingly (e.g. [16, 28, 82, 101, 138].

Empathic cultural-aware agents that consider social values and norms come with requirements and challenges that put those of current empathic agents to the test:

1. *Explainability*. They explain themselves on the behavioral and motivational level. This requirement is mandatory since empathy is a collaborative process that requires both partners to disclose (private) information to establish a required level of trust (Sect. 2.3).
2. *Observing*. They observe the human–dialog partner on the level of social signals, including voice. Other technological sensors can be added (e.g., pulse or heart-beat). They detect important patterns and sequences of social signals in interaction (e.g., smile, facial expression, gaze behavior, gestures, posture, or physiological values).
3. *Interpreting and Simulating*. They understand the meaning of utterances together with using interpreted behavior and detected social signals (multimodality). For the interpretation of behavior, different knowledge is needed that cover behavioral norms and values for the culture (e.g., Western, Eastern), possible group affiliation(s) (e.g., scouts, researchers, workers), and individual characteristics (e.g., personality). Such agents must consider the social hierarchy (e.g., status), situational (e.g., home or work environment), and relational context (e.g., family member, work colleague, or stranger) too. Moreover, knowledge about internal (subconscious) processes related to mental states and related observable behavior is mandatory. Mainly, a model of emotions distinguished between external (observable) emotions/social signals and internal emotions is required. A focus should be on the processes of *intrapersonal* emotion regulation, coping, and display rules. These processes might be influenced/shaped by culture, group affiliation, and even by family or individual values and norms. They simulate possible mental states of the dialog partner based on possible representations of goals, motivations, and wishes that can be put in relation to the interpreted behavior and internal processes and emotions.
4. *Acting Emphatically*. They (inter)act empathically, respecting cultural, behavioral values and norms of groups and individuals. They show social-communicative abilities such as interpersonal emotion regulation, social mimicry, display rules, and emotional contagion. Therefore, the cultural-aware agent needs

a representation of the dialog partner's motivation, goals (relevant mental states) that use a model of cultural values and norms, and model values and norms of groups and individuals.

5. *Adaptability*. They can adapt to individuals on various levels considering the agent's role and status. This process starts by respecting individual aspects (e.g., physiological, like the level of hearing or linguistically, like the dialect or idiolect, and cognitively, like the level of cognitive resources). Therefore, they should learn individual characteristics, values, and norms and their relations (1) to internal representations such as motivation, goals, and wishes and (2) to behavioral aspects. The ability to adapt contains the ability to react and to address misinterpretations/simulations, giving apologies for them (e.g., [97]). For them, processes for reflecting and discussing interpretations and learn new values and norms are necessary.

Predictive Empathy

The described cultural-dependent abilities, such as explanation, adaption, and the simulation of possible mental states, including specific internal (subconscious) processed based on a more general model of emotions, are beyond current approaches of empathic agents and demand for a broader concept of empathy that integrates these abilities.

In this work's scope, we define empathy as the ability to simulate how others subjectively experience a situation and how they regulate elicited emotions. We differentiate between: (1) Empathic mentalization and observation (Sect. 2.4.4) with empathy values (Sect. 2.4.3), and (2) Empathic behavior with cultural behavior values (Sect. 2.4.5). Note, that conceptually, empathy is different from empathic behavior. Empathic behavior results from empathic mentalization, observation, using a model of motivations and goals, and empathy/behavior values. The latter is defined by culture, societies, groups, and individuals.

Empathic cultural-aware agents rely on the process of simulation possible internal states of others and compare them with observable behavior considering cultural values and norms. This approach can be seen as a high-level process of prediction. There is scientific evidence of human sensory systems (e.g., vision, audition) could be described as predictive processes [13, 313ff]). These simulations contain, among others, bodily configurations, sights, smells, tastes, touches, and even action tendencies. Simulations are compared continuously with actual sensory input. If they do not match, errors are resolved, and then new predictions and new simulations are created (Fig. 2.4).

Concerning simulating others' internal states related to others, the is a vast challenge to realize computational empathy as a predictive process(es). This approach will interweave affective and cognitive aspects in one model to generate a more general computational model of empathy that can simulate internal mental states depending on cultural values. Based on them, empathic reactions can be generated that reflect the cultural norms of empathic behavior. Functionality-wise, the predic-

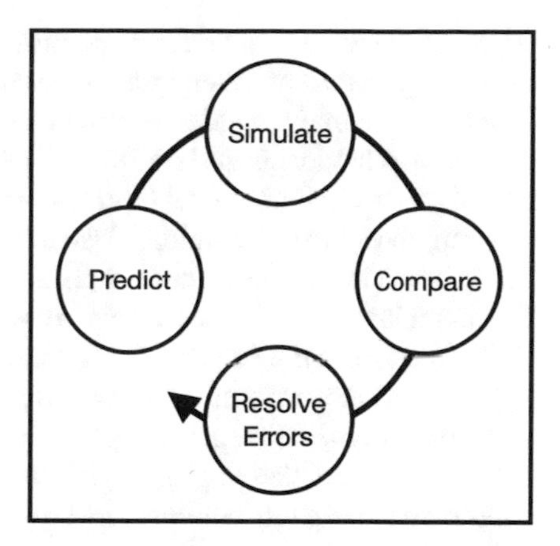

Fig. 2.4 Prediction loop

tive empathy process can be related to the functional architecture of human empathy by [34], which describe three core skills for human empathy: (1) ability to share the other person's feelings, (2) cognitive ability to simulate what another person is feeling, and (3) a "socially beneficial" intention to respond compassionately to a person's negative emotion.

Beyond Empathy

Besides this process view of empathy, it is crucial to consider knowledge models that represent, e.g., different cultural communication styles, different cultural values, individual values, and relevant persons of a person's life. It must be noted that cultural and individual values are expressed through fashion, literature, design, music, and other, e.g., [25, 31]. Thus, they might be used by agents to convey common values, which can be considered a "static version of empathy."

2.4.2 Computational Models of Emotions and Empathic Behavior

Computational models of empathy and emotions can be theory-driven or data-driven and can be verified by empirical means (Sect. 2.6). The theory-driven approach uses observations, theories, and models from various disciplines, e.g., psychology, social science, and creates computational models, as it has been done in various related approaches for the related topic of computational models of affect and empathy since the 1980s [102]. For overviews, check [83, 111]. Concerning the emotional behavior of agents that consider cultural values, first systems have emerged [43, 107]. Since such models are based on explicit symbolic modeling; their functionality is easily explainable by relying on the respective theories. The data-driven approaches focus on the replication of behavioral aspects of human emotional or empathic behavior. Therefore, people are video recorded in situations in which they show emotional or empathic behavior. Based on this data and annotations that describe the data concerning a specific emotional or empathic behavior in a specific situation, a machine learning model (ML) is trained. Usually, a large amount of video material is needed. These models can be used to replicate such behaviors for the learned situations. The explainability of such models is complicated and is an emerging research trend [79]. Apart from the fact that the replicated behavior is generated automatically, this approach cannot inform about underlying motivations and goals, which is necessary for the type of empathic agents described here. Moreover, the function of the ML-model is challenging to explain in terms of human ways of explanations. To overcome those short-comings, first, hybrid-approaches have been created [79, 90].

2.4.3 Empathic Values, Culture, and Ethics

For empathy concepts, culture plays a central role. Culture defines a frame for those who are part of that culture and provides security and explanations. Culture serves two functions: (1) integrative, which provide individuals with a sense of identity, and (2) functional, e.g., rules for survival and welfare, establishing self-worthiness and belonging. These functions enable members to maintain social and behavioral consistency. They form recognizable cultural patterns are define social interaction and integration [24]. Cultures are often described with the concepts of values and norms. They represent agreed-on views of what is right and desirable, and they denote cultural ideals [115].

"Cultural value emphases shape and justify individual and group beliefs, actions, and goals. Institutional arrangements and policies, norms, and everyday practices express underlying cultural value emphases in societies. For example, a cultural value emphasis on success and ambition may be reflected in and promote highly competitive economic systems, confrontational legal systems, and child-rearing practices that pressure children to achieve." [115, p. 139]. Hence, some norms that are related to empathy might differ for generations. Norms of different generations within a society can be seen as "different" cultures [38, 44]. Some norms are shared over generations, but new values are integrated, cf. generation gap/conflict [14, 137]. There are smaller societal entities, like cohorts or groups, that also can be distinguished by a different set of values, and norms [2]. In the western world, a valid empathic response could be mirroring the other person has expressed emotions or being in sync (congruent) with the other person's (inner) emotional state. However, this can vary between specific groups and individuals, and the individual's brain structure might even reflect this [8].

In social sciences, culture is mostly represented with dichotomous (dichotomous) dimensions representing opposite value orientations, such as the cultural dimension autonomy—embeddedness. Other types of cultural representation relying on cate-

Fig. 2.5 Cultural dimensions: prototypical structure [115, p. 142]

gorical methods representing independent categories [35]. We focus on dimensional representations of culture. A point on that dimension describes to what extent people are autonomous versus embedded in their groups. "In autonomy cultures, people are viewed as autonomous, bounded entities. They should cultivate and express their preferences, feelings, ideas, and abilities, and find meaning in their uniqueness." [115, p. 140]. There are many theory-based approaches to model culture. Focusing on organizational and work values [60] describes a highly recognized theory of culture, values, and related behaviors using the five dimensions power distance (cf. hierarchy), identity (cf. autonomy), gender, uncertainty avoidance, and long-term orientation. House and colleagues provide a more fine-grained model that describes work and organizational cultural values with nine dimensions, the five from Hofstede plus performance orientation, assertiveness, institutional collectivism, and humane orientation [63]. Other theories of cultures highlight specific characteristics, such as, e.g., the level of explicitness in conversations [51], or on emotion regulation and emotional behavior related to the cultural background [3] or to related to value dimensions [85–87]. Their work is very relevant in the context of culturally appropriate emotional behavior, including emotion regulation (Sect. 2.4.5).

A theory of culture that focuses on underlying motivation, such as social attitudes and motivation, which is relevant for empathic agents, is designed by Schwartz [115]. There, cultural values are organized with three cultural dimensions: (1) embeddedness versus autonomy (intellectual vs. affective), (2) hierarchy versus egalitarianism, and (3) mastery versus harmony, and with seven cultural value orientations (Fig. 2.5). Based on an empirical mapping of 76 national cultures, seven transnational cultural groupings are distinguished: West European, English-speaking, Latin American, East European, South Asian, Confucian influenced, and African and Middle Eastern. The theory was evaluated with the help of the Schwartz Value Survey (SVS) questionnaire and the (European Social Survey) ESS questionnaire (includes a short version of the Portrait Values Questionnaire (PVQ)) to validate the seven cultural value orientations. The data analysis suggests that they support explanations of fundamental social phenomena uniquely and hence might serve as a good starting point for modeling cultural values, attitudes, and norms for empathic agents. The analysis [115, p. 153ff] allows explanations of implications of societal characteristics like, e.g., the American culture shows values that support mastery and affective autonomy and values that do not emphasize harmony. The culture in Singapore, e.g., shows values that support hierarchy but not egalitarianism and intellectual autonomy.

While these findings help organize a knowledge base of cultural values and give empathic agents a starting point, individual persons' values have to be collected to let the agent adapt to the person's individual "cultural" profile. A starting point could be the operationalization of the value questionnaires from, e.g., Schwarz or Inglehart and Baker [66, 115, 116]. Schwartz provides an overview of how fundamental cultural values can be represented [115, p. 143]. Values …

- *are beliefs*. Beliefs are linked to affective experience, and hence are connected to the concept of empathy (Sect. 2.4.4)
- *refer to desirable goals*. Goals motivate action.
- *transcend specific actions and situations*. "(e.g., obedience and honesty are values that are relevant at work or in school, in sports, business, and politics, with family, friends, or strangers). This feature distinguishes values from narrower concepts like norms and attitudes that usually refer to specific actions, objects, or situations."
- *serve as standards and criteria*. They influence the appraisal of actions, events, objects, and people. This is the connection to a cognitive appraisal component that can be realized within empathic agents (Sect. 2.4.4).
- *are ordered by importance in a system of priorities*. "This hierarchical feature also distinguishes values from norms and attitudes."
- *have relative importance that guides action*. "The tradeoff among relevant, competing values is what guides attitudes and behaviors."

Technological Approaches

It seems that the representation of cultural values, the underlying representation of motivation, and attitudes for empathic agents can be realized at best with ontological representations. In the area of HCI, Blanchard and colleagues provide a first ontological approach to how affect and cultural values can be represented computationally [17]. Such approaches have to be extended by concepts of relationship and personal development, for example, discussed by Bickard [15].

Empathy comes with additional requirements, such as simulating how another person might have experienced a situation based on individual and cultural values and experiences. Empathic agents should simulate possible interpretations of how others might experience (emotionally) a situation. Those simulations can be described with the help of a BDI-model [106] (Sect. 2.4.4) where facts and rules are used first to simulate possible situations and then simulate possible interpretations.

Some Ethics and Enculturation

An in-depth discussion of cultural values and how they are represented in societies is beyond this book's scope. However, the philosophical perspective about cultural values, their change over time, and their connection to empathy should be considered by any empathic agent researchers.

For the relevant area of care, a starting point is a work from psychologist Mary Jeanne Larrabee, who discusses care ethics from an interdisciplinary perspective [74]. Another starting point could be the work from the philosopher of education Nel Noddings, who discusses ethics of care from the perspective of human relationships [96]. It could be stated that in this context, affect and empathy serve as a common language and action motivator. Both are strongly influenced by social, cultural, and familial value [120, chap. 2]. Values influence how we react empathically, e.g., racial favoritism is not "naturally given"; it is a cultural value. In the context of empathy and caring, Slote argues that empathy should be a central part of the process of caring [120, p. 58f]. Moreover, we can adopt this view to how we want empathic agents to act as care assistants. He points out that concerns for wellbeing and respect are in danger in situations of paternalism, e.g., when someone acts against another person for her or his good. One crucial question is, when is paternalism justified? He argues that if caring includes empathy (Sect. 2.4.4), the ethics of caring would account for the cultural values of respect and autonomy. As mentioned before, he empathizes that the process of empathy requires the ability to separate one owns desires, goals, and affect from others'. This aspect might be an aspect that can be easily achieved by an empathic agent. Within the context of autonomy and respect, it seems to be important how to handle cognitive and bodily restrictions (e.g., [36]). Other relevant situations within that context might be how to handle loneliness.

Using technology implies that other (foreign) values might "enter" the cultural group's set of values. In that context, an important research topic is to investigate ways how empathic agents can be introduced into a cultural group. Spiekermann discusses how the values like autonomy, health, safety, security, privacy, trust (Sect. 2.3), friendship, as well as dignity and respect can be incorporated into a value-based IT design [121]. While Schwartz and his research colleagues (Sect. 2.4.3) investigates cultural values from the perspective of Social Sciences, Spiekermann focuses on the psychological view and the economics view, which provides an additional perspective to cultural values and norms.

2.4.4 Emotions, Empathic Mentalization

Empathy is related to emotions. For the following explanations, we rely on the theory of emotion by Moser and von Zeppelin [92] and is realized in a computational model that relates appraisal, intrapersonal emotion regulation, and social signal interpretation [47]. It differentiates between communicative emotions, structural emotions, and situational emotions. Whereas communicative emotions can be seen as externally observable emotional expressions (e.g., happiness, surprise) and related social signals (e.g., smiling, nodding), structural and situational emotions denote internal emotions that reflect individual subjective experiences. Within that context, structural emotions represent information about the appraisal of oneself and are related to the self-image, e.g., shame, pride, and gratitude. Situational emotions reflect the level of security and represent information linked to a topic or situation that has been experienced. More specifically, such emotions as, e.g., fear or distress, reflect that the situation comes with unforeseen or unbearable requirements. If a situation addresses social skills or relations, the emotions shame or pride might be linked.Another essential requirement for the simulation of emotions (as subjective experiences of others) and their empathic management is the representation of interpersonal emotion regulation [50].

2.4.5 Empathic Behavior and Culture

What can be considered as empathic behavior or empathic responses to another person's situation can be related to values and norms of individuals and groups (societies) and their culture [7, 12, 59, 60, 85, 87, 94, 95] (Sect. 2.4.3). As mentioned above, the dimension of individualism can be used to distinguish cultures. How the type of individualism influences empathic behavior and empathic mentalization has been researched recently. Atkins and colleagues report that there are diverging patterns of empathic response between cultural groups [7]. People from the Western cultural context usually show other-focused emotional responses and turn their attention towards the person in distress, whereas people from the Eastern cultural context do otherwise. In response to others' negative experiences, the emotional responses of people from the Eastern culture are more focused on themselves and regulating their own emotions than people from Western culture. Chang and Bemark point out that in cross-cultural counseling, one has especially to consider aspects of how to behave empathically in the light of traditional and cultural values [24]. In this context, they point out that the importance of the level of individuality might be perceived differently. In individualistic societies, it may be a high value, whereas, in collectivist societies, the boundaries of separateness and personal and spiritual independence may be less honored.

Emotional Expressions

Many studies report that the expression of some (basic) emotions is universal over the world. However, this view seems not to be the case, and hence, cannot be used to created cross-cultural rules for emotional expression and empathic behavior. As research from different disciplines suggests, each individual constructs its own perception of emotional expression, which is related to one own's concept of emotions [13, p. 50], that has been learned in early childhood [48] [46, p. 153ff] (Sect. 2.4.4). As a result, empathic agents must actively learn or must have at least a representation of what is considered to be empathic behavior concerning the individual itself, the groups it is a member of. The latter could be a starting point (cf. research from Atkins and colleagues [7]) for modeling the empathic behavior of empathic agents. In this context, the use of emotional display rules [41] to show situational appropriate emotional expressions seems a first approach, which could be operationalized in automatic behavior of empathic agents [5]. As a next step, these agents must learn/adapt to individual or group accepted empathic behavior.

Functions of Emotions, Social Signals, and Caressing Actions

Emotions have different functions [18]. One function is the internal information about the current situation concerning oneself. Other external functions of expressed emotions are the (1) illustration of utterances and (2) information concerning regulating interpersonal relationships. All three functions of emotions are essential for empathy, and empathic behavior [9].

Other observable behavior functions related to empathy is mimicry behavior, synchronicity and behavioral attunement [55, 73], and verbal and non-verbal expressions of interpersonal emotion regulation [50]. Caressing actions and supporting gestures, e.g., touching one other or helping maintain a firm stand (to prevent falling), can be considered relevant for empathic behavior.

Technological Approaches

In terms of technical realization, the creation go behavior for such agents is either theory-based, data-driven or a combination of both [79, 90]. A theory-based representation is a first starting point that can be used by empathic agents to show and observe culture-dependent behaviors. For empathic agents, a meticulously fine-grained representation of norms and values, motivation, attitudes, and their relation to multimodal empathic behavior models is needed. As shown above, most theories on culture are based on dimensional traits that can be transferred in computational models with an n-dimensional numerical representation of cultural norms and values, like André has pointed out for intercultural emotional agents [5]. This approach has been successfully combined with a data-driven approach that comes with multimodal recordings of existing cultures. A Bayesian Network is used to model culture-dependent non-verbal behaviors for a given conversation in three different scenarios [79]. Al-Saleh and Romano provide a review of appropriate cultural behavior of virtual agents [4].

Paiva et al. give an overview of existing approaches of computational models for virtual and physical agents (17) and distinguish agents that evoke empathy in users and agents that are designed to be empathic. We focus on the latter group. The authors argue that most approaches follow the perception-action hypotheses (automatically explicit an empathic response). This considers observed emotions, social relationships, the situation, the observer's context and features, and the emphatic response.

2.5 Research Agenda

As mentioned in the above sections, a variety of models for modeling trust and empathy are proposed. However, evaluating such models in the real field is a remaining problem because introducing robots in the daily life of aged people has just begun in recent. In this section, we indicate some typical problems caused by introducing robots in the aging society.

2.5.1 Problems Caused by Recording

One of the most significant issues is caused by sensors that robots have. Robots generally have cameras or microphones to interact with the environment; however, it often conflicts with aged people's privacy. Smart speaker devices, which are recently launched by several companies, face this problem because they can record audio anytime; thus, the psychological resistance to record audio in daily life should be researched. If the device has cameras in addition to microphones, the risk of infringing on privacy will increase. Developers of such devices should be careful of privacy and user tolerances to set access of equipment to the environment in the daily life of aged people.

The problems caused by recording happen for any generation; however, problems will be more complicated, especially for the aged people because the recorded data is mainly accessed by surrounding people of the aged people, who need to watch them. Access to the recorded audio or video in such a case is a critical case. In many cases, families or doctors will use such devices to follow aged people's daily lives; however, there is a trade-off between the accessibility to the recorded results and invasion of the privacy of the aged people. What kinds of controls are reposed to the aged people themselves and the surrounding people of the aged people who need to watch them? We still need a discussion and a psychological survey to find a balanced point.

Trust from the aged person to the robot is an essential factor that changes the balance point. If the user has enough trust in the system, the user will entrust information access to the robot. However, distrust of the robot will change their minds to limit the available devices. Empathy is an important factor in improving trust.

2.5.2 Empathic Behavior of Robots

Empathic behavior will solve problems caused in communication between aged people and the system by building a healthy relationship. If people have empathy for another person, trust in the person also appears. Empathy and trust complement each other as the two wheels of a cart; however, modeling both is still an underlying problem.

Several existing studies tried to model empathy by using emotion elicitation [78] or attentive listening [142]. It is reported that their resultant models improved the empathy of users; however, relation to trust is not researched. Some recent studies reported that proactive behavior of the system would increase user activities and empathy to the system [91, 117, 140]. However, these works only measure a ratio of increased user utterances and do not analyze dialogue data from the viewpoint of empathy and trust. Quantitative analysis is required between robot behaviors and empathy to the robot of users. Methods to measure empathy or trust will be discussed in Sect. 2.6.

Relations between physical issues of robots and empathy or trust are also an important research topic. If the robot has a huggable surface like a piece of fabric, they contribute to empathy [93] or trust [124] of users. Appearances and expressions of humanoid robots also significantly impact empathy or trust from users [49]. These phenomena are only researched on typical robots; thus, we need to build a theory to design empathic robots, which will be trusted by aged people.

2.5.3 Explainability

Explainability is also an issue, system notification to the aged people. Suppose it is essential to record aged people's daily lives to prevent problems, e.g., problems caused by dementia. Is it acceptable to record their everyday life without notifying that to them? When the number of autonomous robot functions is increased, controlling these behaviors also will be a problem. If we respect aged people's rights, they should have the right to control the robot's behavior fully; however, it often conflicts with the purpose of the robot, watching the aged people. We also need to consider the aged person's judgment skills to give them the rights to turn on/off functions. Systems in the aged society should have the ability to persuade the aged people, watched by the systems.

When we give rights to access robot controls to aged persons, the description level to be used for the aged person is also important. The system should have transparency about what is recorded for which purpose; however, it is hard for systems to describe the detail in an understandable explanation for aged people. In that situation, will it be allowed for the system to describe the situation by using not accurate descriptions or deceptive descriptions? For example, there is a standard test to check the dementia of aged people, and the robot will conduct such tests [141]. However, some aged people know the test and decline to have the test. Is it allowed for the robot to use deceptive descriptions or not-true descriptions in such situations? Probably it depends on the types of robot assistance; however, accessing the acceptance in the balance of the obligation of the system and the right of aged person should be researched. The balanced point will change by the social situation and roles of robots in society.

The other importance of explainability stands on overtrust, which is mentioned in the above sections. People trust the behaviors of industrial products, including robots; however, overtrust will cause some problems if the robot does not have enough skills. This situation will happen in the early stage of robot introduction to society. Explaining current situations and the recognition results of the robot are important to prevent this problem. Some related works try to bridge robot behaviors and explanations [105, 125, 139]. Considering the uncertainty of robots in such works will be important to improve the explainability.

The explainability of robots also will solve the problem of conflict between the user and the system. Some existing works tackled preventing physical contacts between the robot and the user [118]; however, conflictions in tasks or older people are not well researched. Aged people tend to be shortsighted due to their physical declination. Considering the declination in confliction modeling is essential.

2.5.4 Conversation Between Robots and Aged People

According to viewpoints mentioned in the above sections, conversations between robots and aged people are essential to prevent problems. However, if we try to implement conversational functions on robots to talk with aged people, many technical challenges are underlying. Automatic speech recognition (ASR) for an aged person's voice is difficult because their speech sounds are small and unclear. Current ASR systems work well if the voice is spoken for dictation; however, recognizing speeches in natural conversation, which are not expressed in systems' awareness, is still challenging. Human-like robots will cause users' speeches, especially if the user has high trust or empathy for the system.

To prevent the misunderstanding in the communication, actions of confirmation and clarification are useful. It sometimes hurt the system's usability; however, the system needs to decide the balance between the misunderstanding risk and the task success of the system. This is a conventional research agenda to build task-oriented dialogue systems [143]. Reinforcement learning is generally used for this problem, and we need to visit them to learn what the research agenda will be when we try to model a conversation between robots and aged people. Initiative and task scenarios are also critical because we can reduce the risk of misunderstanding by setting up the task scenario and assumed speaker in each situation. If the robot in a situation where the misunderstanding is serious, the robot should use scenarios to have the dialogue initiative prevent problems. On the other hand, the robot can give the dialogue initiative to the aged person when they try to improve empathy or trust by using attentive listening.

2.6 Evaluation Methods

The present section will review the evaluation methods of trust and empathy since we need to evaluate the created system. This section introduces several items that were developed to measure empathy and trust by questionnaires, and we describe lists and the procedure to create them by referring to some examples. Also, some papers applied them to the context of human–computer interaction. We can see that evaluation was done in cognitive, affective and their sub-components separately. Further study will be needed to investigate cultural difference of empathy and trust.

2.6.1 Measuring Empathy

The term empathy refers to both the sharing of emotions between individuals and the adopting of another's point of view [135]. Empathy is a basic human capacity that serves to regulate relationships, supporting collaboration and group cohesion. It refers to the ability to respond affectively to emotions in others, aiming at reacting adaptively to another's needs [109]. One theory of empathic responses distinguishes between two components of empathy: affective empathy and cognitive empathy. The affective component of empathy refers to individuals' emotional reactions in response to another person's feelings that typically mirror the other person's feelings or are congruent with his or her emotional state. The two most commonly examined indices of affective empathy are personal distress and empathic concern. Here, personal distress has been defined as an aversive response to witnessing someone else's negative emotional state and is conceptualized as a self-focused emotional response associated with motivation to attenuate one's own aversive feelings. Empathic concern is usually conceptualized as an other-focused emotional response and is associated with attention turning toward the person in distress (sympathy). In contrast, the cognitive component of empathy refers to accurately recognizing another person's thoughts and feelings and is mainly focused on the underlying cognitive processes such as perspective-taking or accurately recognizing another's emotions. The most commonly examined index of cognitive empathy is empathic accuracy that refers to individuals' successful inferences of targets' feelings [64, 69].

Self-report measures are validated approaches to measuring empathy in human–human interaction, and adapted to human–computer interaction [101]. A study [6] examined cultural effects for empathic responses to physical and social pain would be a good starting point to evaluate affective and cognitive empathy. They utilized several items. Following is the list of the measures.

Interpersonal Reactivity Index (IRI) the instrument developed by Davis [32] for self-assessment of the own ability of empathy and consists of four seven-item subscales: fantasy scale, perspective-taking scale, empathic concern scale, and personal distress scale. This was typically used to assess the cognitive component of empathy.

Affect rating Participants were instructed to provide a continuous report of their positive and negative affective state as they watched each video using a rating dial.

Positive and Negative Affect Schedule (PANAS) 5 points Likert scale (1: very slightly or not at all to 5: extremely) to reflect their feelings when they described their experiences [136]. Absolute difference scores between each PANAS emotion score reported by the targets in the videos and those reported by the participants were calculated in order to rate empathic accuracy.

Emotional Response Questionnaire (ERQ) the instrument consists of six emotional adjectives (compassionate, sympathetic, moved, tender, warm, and softhearted) [27]. Adjectives included as possible measures of personal distress were alarmed, perturbed, disconcerted, bothered, irritated, disturbed, worried, uneasy, distressed, troubled, upset, anxious, and grieved. Two other adjectives, intent and intrigued, were included to provide some adjectives relevant to the first broadcast, an informative announcement. Adjectives included as possible measures of empathic concern were moved, softhearted, sorrowed, touched, empathic, warm, concerned, and compassionate. This is used for rating empathic concern.

Other studies have proposed items related to evaluating empathy. We summarize them as follows. Several empathy questionnaires were summarized in [122]. The aim of the study was to identify common factors among different conceptions of empathy as operationalized by published relating measures such as the IRI, the Autism Quotient (AQ) [11], the Reading the Mind in the Eyes Task (MIE; [112]), Interpersonal Perception Task [65], and Empathy Quotient [75]. They finally constructed the Toronto Empathy Questionnaire (TEQ) through a series of studies. TEQ consists of 16 items to assess human empathy.

A questionnaire of cognitive and affective empathy (QCAE) was developed and validated in [108]. The study prepared a total of 65 items of measuring cognitive (29 items) and affective (36 items) empathy which was derived from the Empathy Quotient [10], the Hogan Empathy Scale [61], the Empathy subscale of the Impulsiveness-Venturesomeness-Empathy Inventory [45], and the IRI [32]. They finally constructed the 31-item version of the QCAE which was validated by the principal component analysis (PCA) and the confirmatory factor analysis (CFA).

In addition to the self-report items, an Empathy Questionnaire (EmQue), a parent questionnaire regarding empathy-related behaviors in young children was developed [109]. The EmQue consists of 20 items, representing three facets of empathy that should be observable in young children: (1) Emotion Contagion, (2) Attention to Others' Feelings, and (3) Prosocial Actions. Parents can rate the degree to which each item, reflecting a type of behavior, applied to their child over the past two months on a 3-point scale (0 = never, 1 = sometimes, 2 = often).

Not only investigating empathy in the context of human–human interaction but in this book, we also consider transferring it to human–computer interaction [101]. There is a survey article investigating empathy in virtual agents and robots. They introduced some of the above self-report questionnaires and finally proposed an empathy questionnaire for social robots which was inspired by [32]. Below are the first two questionnaires.

- <agent-name> can have tender and concerned feelings for people less fortunate than himself.
- Sometimes <agent-name> found it difficult to see things from my point of view.

Another study is aimed to evaluate empathy in an artificial interactive system. They proposed system-level and feature-level evaluations for empathy systems in a systematic way [99].

Here we introduce examples of how we apply the measures to cultural differences in empathy. Evidence of cultural differences in the interpretation of the self and interpersonal relationships suggests that empathic responses to others' emotional states would change as a function of cultural background. But, there is limited empirical research conducted to examine the role of culture in empathy. There is a study that East Asian and White British participants differ in both affective and cognitive components of their empathic reactions in response to someone else's pain [6]. Trommsdorff et al. [131] found that children from other-oriented cultural groups (Indonesia and Malaysia) showed more personal distress than did children from individual-oriented cultural groups (Germany and Israel). However, they did not observe any cultural group differences in empathic concern. Cassels et al. [22] examined cultural differences in empathy focusing on individual differences in empathic concern and personal distress among East Asian and European Canadian young adults by using IRI [32]. They found that Westerners reported more empathic concern, but less personal distress than did Easterners.

2.6.2 *Measuring Trust*

In this subsection, we describe some self-report measures of trust. We consider both situations: human–human interaction and human–computer interaction. Firstly, there has been a study conducted an analysis of associated words to trust and distrust [67]. The cluster analysis was performed to visualize words related to human–human trusts and human–machine trusts. This result could be used to imagine the word trust in both situations.

Proactive decisions to the developed system should match the users' preferences to maintain the users' trust in the system. Wrong decisions could negatively influence the users' acceptance of a system and at worst could make them abandon the system.

In a study [53], a trust-based model, called User Trust Model (UTM), for automatic decision-making was proposed. The following categories have formed the basis of their UTM [33]:

- Comfort of Use ("The system should be easy to handle")
- Transparency ("I need to understand what the system is doing")
- Controllability ("I want to be in control of the system's actions")
- Privacy ("The system should neither ask for nor reveal private information")
- Reliability ("The system should run in a stable manner")
- Security ("The system should safely transfer data")
- Credibility ("The system should have been recommended by others")
- Seriousness ("The system should have a professional appearance").

Furthermore, we distinguished at the previous section between a static view of trust—the inherent *trustworthiness* of a system—and trust as a dynamic relationship with a changing value. We could see this as a parallel distinction to that between mood and dynamic affective state in that the two will interact. Thus a dynamic failure of trust may impact a system with a high level of trustworthiness less than one with low trustworthiness, though repeating such lapses might lead to a drop in overall perceived trustworthiness.

Measurement must deal with both aspects: the more static assessment of the perceived trustworthiness of a system, and the changes in trust over time. The mechanisms are the same as those used for affective state: questionnaires, social signal processing, performative cues, and physiological signals.

Of these methods, the most straightforward—questionnaires—is also strongly slanted towards the static aspects of trust. One can implement pre- and post- questionnaires around an interaction but the within-interaction dynamics are likely to be largely missed, short of deliberately engineered breakdowns of trust. In the wild, and especially with the elderly, questionnaires are of limited use, though where feasible, interaction diaries are likely to be useful. However, they are a useful experimental tool at a time when much has yet to be learned about trust.

A questionnaire implicitly codes aspects of a model in deciding what questions to include. Schaefer [114] has forty questions based on the model of Fig. 2.6 and Hancock's triadic model of Fig. 2.2.

This uses a distinction between cognitive and affective trust earlier discussed in [80], a parallel to that between cognitive and affective empathy. They define the five constructs (see Fig. 2.7 for expected relationships (E1-E7) between the five constructs and trust) and their corresponding five items produced from the research which are shown below:

ÆŠanot

1. Perceived reliability

   - The system always provides the advice I require to make my decision.
   - The system performs reliably.

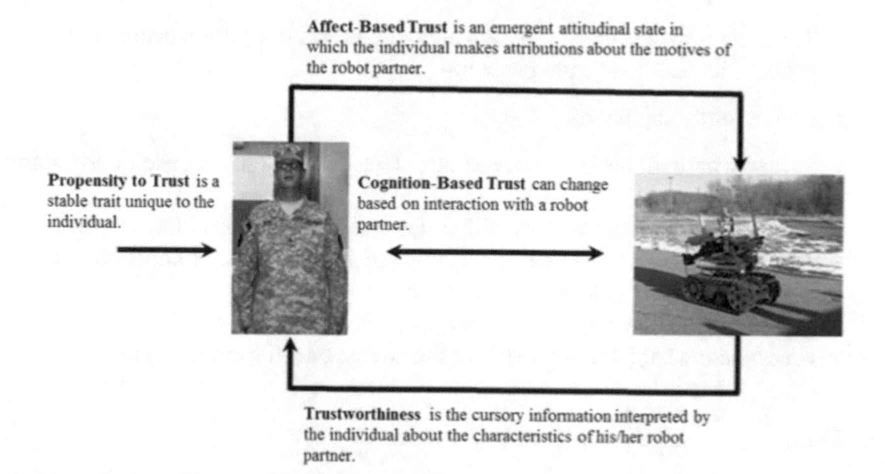

Fig. 2.6 The model used by Shaeffer's questionnaire, see: [114]

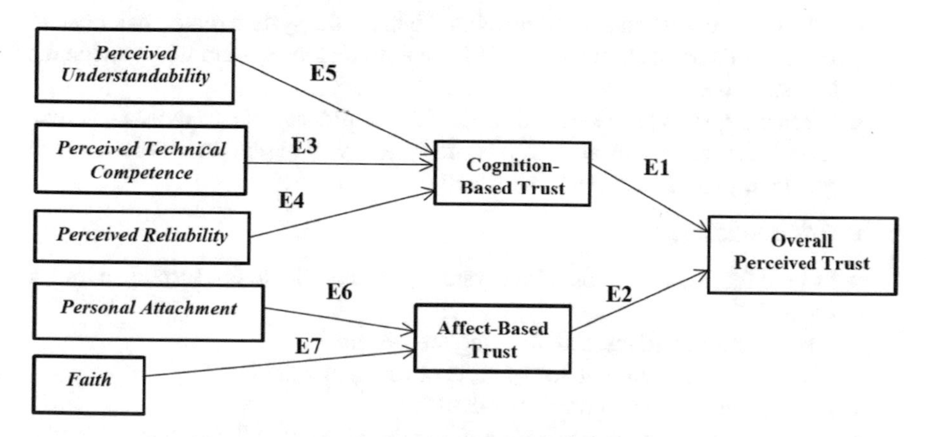

Fig. 2.7 Model of human–computer trust components, see: [80]

- The system responds the same way under the same conditions at different times.
- I can rely on the system to function properly.
- The system analyzes problems consistently.

2. Perceived technical competence

   - The system uses appropriate methods to reach decisions.
   - The system has sound knowledge about this type of problem built into it.
   - The advice the system produces is as good as that which a highly competent person could produce.
   - The system correctly uses the information I enter.

- The system makes use of all the knowledge and information available to it to produce its solution to the problem.

3. Perceived understandability

- I know what will happen the next time I use the system because I understand how it behaves.
- I understand how the system will assist me with decisions I have to make.
- Although I may not know exactly how the system works, I know how to use it to make decisions about the problem.
- It is easy to follow what the system does.
- I recognize what I should do to get the advice I need from the system the next time I use it.

4. Faith

- I believe advice from the system even when I don't know for certain that it is correct.
- When I am uncertain about a decision I believe the system rather than myself.
- If I am not sure about a decision, I have faith that the system will provide the best solution.
- When the system gives unusual advice I am confident that the advice is correct.
- Even if I have no reason to expect the system will be able to solve a difficult problem, I still feel certain that it will.

5. Personal attachment

- I would feel a sense of loss if the system was unavailable and I could no longer use it.
- I feel a sense of attachment to using the system.
- I find the system suitable to my style of decision making.
- I like using the system for decision making.
- I have a personal preference for making decisions with the system.

The use of social signal processing—of which physiological signals are effectively a subset—to detect levels of trust, is a so far little-explored field. The problems of relating an observed signal from a human interaction partner to the social concept of trust are arguably even more difficult than relating it to affect. The experiments cited above that demonstrated overtrust were in general using performative measures that were an integral part of the experimental design. This made them much less ambiguous than social signals, but again, in the wild, they are likely to be much less well-defined and thus much less useful. It is clear that much work remains to be carried out.

2.7 Conclusion

This section has focused on the concepts of trust and empathy and discussed especially their modeling in human–robot interactions in the context of challenges that modern society faces due to demographic, environmental, and societal changes. With the goal of an empathic, interactive agent, the section summarizes current views of trust and empathy. It provides prospects of such agents as service providers for the new interactive applications.

The section defines empathy and trust as having two separate components, cognitive and affective components, and evaluates the concepts from these view-points. Moreover, empathy is regarded as a collaborative process that requires explainability and disclosure of (private) information and is hence connected to trust. As for explainability, hybrid (theory and data-driven) modeling approaches are needed. This involves a simulation of personal experiences and a matching with observable social signals. This process resembles predictive processes that can be found in the human brain.

Finally, cultural aspects need to be taken into account. Empathic cultural-aware agents should be adaptive and must have the ability to learn individual values and norms.

References

1. Abd, M., Gonzalez, I., Nojoumian, M., Engeberg, E.: Impacts of robot assistant performance on human trust, satisfaction, and frustration. In: Morality and Social Trust in Autonomous Robots, RSS (2017)
2. Abrams, D., Wetherell, M., Cochrane, S., Hogg, M.A., Turner, J.C.: Knowing what to think by knowing who you are: self-categorization and the nature of norm formation, conformity and group polarization. Br. J. Soc. Psychol. **29**(2), 97–119 (1990)
3. Allwood, J.: Intercultural communication. Pap. Anthropol. Linguist. **12**, 1–25 (1985)
4. AlSaleh, M., Romano, D.: Culturally appropriate behavior in virtual agents: a review. In: Proceedings of the AAAI Conference on Artificial Intelligence and Interactive Digital Entertainment, vol. 11 (2015)
5. André, E.: Preparing emotional agents for intercultural communication. The Oxford Handbook of Affective Computing, pp. 532–551. Oxford University Press, London (2015)
6. Atkins, D., Uskul, A.K., Cooper, N.R.: Culture shapes empathic responses to physical and social pain. Emotion **16**(5), 587–601 (2016)
7. Atkins, D., Uskul, A.K., Cooper, N.R.: Culture shapes empathic responses to physical and social pain. Emotion **16**(5), 587 (2016)
8. Banissy, M.J., Kanai, R., Walsh, V., Rees, G.: Inter-individual differences in empathy are reflected in human brain structure. Neuroimage **62**(3), 2034–2039 (2012)
9. Bänninger-Huber, E.: Mimik Übertragung Interaktion: Die Untersuchung Affektiver Prozesse in der Psychotherapie. Huber (1996)
10. Baron-Cohen, S., Wheelwright, S.: The empathy quotient: an investigation of adults with asperger syndrome or high functioning autism, and normal sex differences. J. Autism Dev. Disord. **34**(2), 163–175 (2004)

11. Baron-Cohen, S., Wheelwright, S., Skinner, R., Martin, J., Clubley, E.: The autism-spectrum quotient (aq): evidence from asperger syndrome/high-functioning autism, malesand females, scientists and mathematicians. J. Autism Dev. Disord. **31**(1), 5–17 (2001)
12. Barr, J.J., Higgins-D'Alessandro, A.: Adolescent empathy and prosocial behavior in the multidimensional context of school culture. J. Genet. Psychol. **168**(3), 231–250 (2007)
13. Barrett, L.F.: How Emotions are Made: The Secret Life of the Brain. Houghton Mifflin Harcourt, Boston (2017)
14. Bengtson, V.L.: The generation gap: a review and typology of social-psychological perspectives. Youth Soc. **2**(1), 7–32 (1970)
15. Bickhard, M.H.: A process ontology for persons and their development. New Ideas Psychol. **30**(1), 107–119 (2012)
16. Bickmore, T.W.: Relational agents: effecting change through human-computer relationships. Ph.D. thesis, Massachusetts Institute of Technology (2003)
17. Blanchard, E.G., Mizoguchi, R., Lajoie, S.P.: Addressing the interplay of culture and affect in HCI: an ontological approach. In: International Conference on Human-Computer Interaction, pp. 575–584. Springer (2009)
18. Bock, A., Huber, E., Benecke, C.: Levels of structural integration and facial expressions of negative emotions. Z. Psychosom. Med. Psychother. **62**(3), 224–238 (2016)
19. Booth, S., Tompkin, J., Pfister, H., Waldo, J., Gajos, K., Nagpal, R.: Piggybacking robots: human-robot overtrust in university dormitory security. In: Proceedings of the 2017 ACM/IEEE International Conference on Human-Robot Interaction, pp. 426–434. ACM (2017)
20. Brave, S., Nass, C., Hutchinson, K.: Computers that care: investigating the effects of orientation of emotion exhibited by an embodied computer agent. Int. J. Hum. Comput. Stud. **62**(2), 161–178 (2005)
21. Cassell, J., Bickmore, T.: External manifestations of trustworthiness in the interface. Commun. ACM **43**(12), 50–56 (2000)
22. Cassels, T., Chan, S., Chung, W.: The role of culture in affective empathy: cultural and bicultural differences. J. Cogn. Cult. **10**, 309–326 (2010). https://doi.org/10.1163/156853710X531203
23. Chen, R., Jankovic, F., Marinsek, N., Foschini, L., Kourtis, L., Signorini, A., Pugh, M., Shen, J., Yaari, R., Maljkovic, V., Sunga, M., Song, H.H., Jung, H.J., Tseng, B., Trister, A.: Developing measures of cognitive impairment in the real world from consumer-grade multimodal sensor streams. In: Proceedings of the 25th ACM SIGKDD International Conference on Knowledge Discovery & Data Mining, KDD '19, pp. 2145–2155. ACM, New York, NY, USA (2019). http://doi.acm.org/10.1145/3292500.3330690
24. Chung, R.C.Y., Bemak, F.: The relationship of culture and empathy in cross-cultural counseling. J. Couns. Dev. **80**(2), 154–159 (2002)
25. Clarke, E., DeNora, T., Vuoskoski, J.: Music, empathy and cultural enderstanding. Phys. Life Rev. **15**, 61–88 (2015)
26. Coeckelbergh, M., Pop, C., Simut, R., Peca, A., Pintea, S., David, D., Vanderborght, B.: A survey of expectations about the role of robots in robot-assisted therapy for children with asd: ethical acceptability, trust, sociability, appearance, and attachment. Sci. Eng. Ethics **22**(1), 47–65 (2016)
27. Coke, J.S., Batson, C.D., McDavis, K.: Empathic mediation of helping: a two-stage model. J. Pers. Soc. Psychol. **36**(7), 752–766 (1978)
28. Conati, C., Maclaren, H.: Empirically building and evaluating a probabilistic model of user affect. User Model. User-Adap. Inter. **19**(3), 267–303 (2009)
29. Cramer, H., Goddijn, J., Wielinga, B., Evers, V.: Effects of (in) accurate empathy and situational valence on attitudes towards robots. In: 2010 5th ACM/IEEE International Conference on Human-Robot Interaction (HRI), pp. 141–142. IEEE (2010)
30. Dautenhahn, K.: Socially intelligent robots: dimensions of human-robot interaction. Philos. Trans. R. Soc. B: Biol. Sci. **362**(1480), 679–704 (2007)
31. Davis, F.: Fashion, Culture, and Identity. University of Chicago Press, Chicago (2013)

32. Davis, M.H.: Measuring individual differences in empathy: evidence for a multidimensional approach. J. Pers. Soc. Psychol. **44**, 113–126 (1983)
33. Davis, M.H.: Personalization of content on public displays driven by the recognition of group context. Ambient. Intell. **7683**, 272–287 (2012)
34. Decety, J., Jackson, P.L.: The functional architecture of human empathy. Behav. Cogn. Neurosci. Rev. **3**(2), 71–100 (2004)
35. Detweiler, R.A.: Culture, category width, and attributions: a model-building approach to the reasons for cultural effects. J. Cross Cult. Psychol. **9**(3), 259–284 (1978)
36. Dilworth-Anderson, P., Gibson, B.E.: The cultural influence of values, norms, meanings, and perceptions in understanding dementia in ethnic minorities. Alzheimer Dis. Assoc. Disord. **16**, S56–S63 (2002)
37. Doney, P.M., Cannon, J.P., Mullen, M.R.: Understanding the influence of national culture on the development of trust. Acad. Manag. Rev. **23**(3), 601–620 (1998)
38. Edmunds, J., Turner, B.S.: Generational Consciousness, Narrative, and Politics. Rowman & Littlefield Publishers, Maryland (2002)
39. Eisenberg, N.: Empathy-related responding and prosocial behaviour. In: Novartis Foundation Symposium, vol. 278, p. 71. Wiley Online Library (2006)
40. Eisenberg, N., Miller, P.A.: The relation of empathy to prosocial and related behaviors. Psychol. Bull. **101**(1), 91 (1987)
41. Ekman, P., Friesen, W.V.: Measuring facial movement. Environ. Psychol. Nonverbal Behav. **1**(1), 56–75 (1976)
42. El Kaliouby, R., Teeters, A., Picard, R.: An exploratory social-emotional prosthetic for autism spectrum disorders, pp. 3–4 (2006). https://doi.org/10.1109/BSN.2006.34
43. Endrass, B., Rehm, M., André, E.: Culture-specific communication management for virtual agents. In: Proceedings of The 8th International Conference on Autonomous Agents and Multiagent Systems - Volume 1, AAMAS '09, pp. 281–287. International Foundation for Autonomous Agents and Multiagent Systems, Richland, SC (2009)
44. Eyerman, R., Turner, B.S.: Outline of a theory of generation. Eur. J. Soc. Theory **1**(1), 91–106 (1998)
45. Eysenck, S.B.G., Eysenck, H.J.: Impulsiveness and venturesomeness: their position in a dimensional system of personality description. Psychol. Rep. **43**(3), 1247–1255 (1978)
46. Fonagy, P., Gergely, G., Jurist, E.L.: Affect Regulation. Routledge, Mentalization and the Development of the Self (2018)
47. Gebhard, P., Schneeberger, T., Baur, T., André, E.: MARSSI: model of appraisal, regulation, and social signal interpretation. In: Proceedings of the 17th International Conference on Autonomous Agents and Multiagent Systems, pp. 497–506 (2018)
48. Gergely, G., Unoka, Z.: Attachment, Affect-Regulation, and Mentalization: The Developmental Origins of the Representational Affective Self, chap. 11. Oxford University Press, Oxford (2013)
49. Glas, D.F., Minato, T., Ishi, C.T., Kawahara, T., Ishiguro, H.: Erica: The erato intelligent conversational android. In: 2016 25th IEEE International Symposium on Robot and Human Interactive Communication (RO-MAN), pp. 22–29. IEEE (2016)
50. Gross, J.J.: Emotion regulation: past, present, future. Cogn. Emot. **13**(5), 551–573 (1999)
51. Hall, E.T., Hall, E.: How Cultures Collide. Psychol. Today **10**(2), 66 (1976)
52. Hamacher, A., Bianchi-Berthouze, N., Pipe, A.G., Eder, K.: Believing in bert: using expressive communication to enhance trust and counteract operational error in physical human-robot interaction. In: 2016 25th IEEE International Symposium on Robot and Human Interactive Communication (RO-MAN), pp. 493–500. IEEE (2016)
53. Hammer, S., Wißner, M., André, E.: Trust-based decision-making for smart and adaptive environments. User Model. User-Adap. Inter. **25**(3), 267–293 (2015)
54. Hancock, P.A., Billings, D.R., Schaefer, K.E., Chen, J.Y., De Visser, E.J., Parasuraman, R.: A meta-analysis of factors affecting trust in human-robot interaction. Hum. Factors **53**(5), 517–527 (2011)

55. Hess, U., Fischer, A.: Emotional mimicry as social regulation. Pers. Soc. Psychol. Rev. **17**(2), 142–157 (2013)
56. Hoffman, M.L.: Empathy, its development and prosocial implications. In: Nebraska Symposium on Motivation, vol. 25, pp. 169–217 (1977)
57. Hoffman, M.L.: Sex differences in empathy and related behaviors. Psychol. Bull. **84**(4), 712 (1977)
58. Hoffman, M.L.: Empathy and moral development: implications for caring and justice. Cambridge University Press, Cambridge (2001)
59. Hofstede, G.: Management control of public and not-for-profit activities. Acc. Organ. Soc. **6**(3), 193–211 (1981)
60. Hofstede, G.: Culture's Consequences: Comparing Values, Behaviors. Sage Publications, Institutions and Organizations Across Nations (2001)
61. Hogan, R.: Development of an empathy scale. J. Consult. Clin. Psychol. **33**(3), 307–316 (1969)
62. Hoque, M.E., Courgeon, M., Martin, J.C., Mutlu, B., Picard, R.W.: Mach: my automated conversation coach. In: Proceedings of the 2013 ACM International Joint Conference on Pervasive and Ubiquitous Computing, UbiComp '13, pp. 697–706. ACM, New York, NY, USA (2013). https://doi.org/10.1145/2493432.2493502
63. House, R.J., Hanges, P.J., Javidan, M., Dorfman, P.W., Gupta, V.: Culture, Leadership, and Organizations: The GLOBE study of 62 Societies. Sage Publications, Thousand Oaks (2004)
64. Ickes, W., Stinson, L., Bissonnette, V., Garcia, S.: Naturalistic social cognition: empathic accuracy in mixed-sex dyads. J. Pers. Soc. Psychol. **59**, 730–742 (1990). https://doi.org/10.1037/0022-3514.59.4.730
65. Iizuka, Y., Patterson, M.L., Matchen, J.C.: Accuracy and confidence on the interpersonal perception task: a Japanese-American comparison. J. Nonverbal Behav. **26**(3), 159–174 (2002). https://doi.org/10.1023/A:1020761332372
66. Inglehart, R., Baker, W.E.: Modernization, cultural change, and the persistence of traditional values. Am. Sociol. Rev., pp. 19–51 (2000)
67. Jian, J.Y., Bisantz, A.M., Drury, C.G.: Foundations for an empirically determined scale of trust in automated systems. Int. J. Cogn. Ergon. **4**(1), 53–71 (2000)
68. Kato, S., Homma, A., Sakuma, T., Nakamura, M.: Detection of mild alzheimer's disease and mild cognitive impairment from elderly speech: binary discrimination using logistic regression. In: 2015 37th Annual International Conference of the IEEE Engineering in Medicine and Biology Society (EMBC), pp. 5569–5572. IEEE (2015)
69. Kraus, M.W., Cote, S., Keltner, D.: Social class, contextualism, and empathic accuracy. Psychol. Sci. **21**(11), 1716–1723 (2010)
70. Kuipers, B.: How can we trust a robot? Commun. ACM **61**(3), 86–95 (2018)
71. Kulms, P., Kopp, S.: The effect of embodiment and competence on trust and cooperation in human–agent interaction. In: International Conference on Intelligent Virtual Agents, pp. 75–84. Springer (2016)
72. Kurdyukova, E., André, E., Leichtenstern, K.: Trust management of ubiquitous multi-display environments. In: Ubiquitous Display Environments, pp. 177–193. Springer (2012)
73. Lakin, J.L., Jefferis, V.E., Cheng, C.M., Chartrand, T.L.: The chameleon effect as social glue: evidence for the evolutionary significance of nonconscious mimicry. J. Nonverbal Behav. **27**(3), 145–162 (2003)
74. Larrabee, M.J.: An Ethic of Care: Feminist and Interdisciplinary Perspectives. Routledge, England (2016)
75. Lawrence, E.J., Shaw, P., Baker, D., Baron-Cohen, S., David, A.S.: Measuring empathy: reliability and validity of the empathy quotient. Psychol. Med. **34**(5), 911–919 (2004)
76. Lee, J.D., See, K.A.: Trust in automation: designing for appropriate reliance. Hum. Factors **46**(1), 50–80 (2004)
77. Lester, J.C., Converse, S.A., Kahler, S.E., Barlow, S.T., Stone, B.A., Bhogal, R.S.: The persona effect: affective impact of animated pedagogical agents. In: Proceedings of the ACM SIGCHI Conference on Human Factors in Computing Systems, pp. 359–366. ACM (1997)

78. Lubis, N., Sakti, S., Yoshino, K., Nakamura, S.: Positive emotion elicitation in chat-based dialogue systems. IEEE/ACM Trans. Audio, Speech, Lang. Process. **27**(4), 866–877 (2019)
79. Lugrin, B., Frommel, J., André, E.: Combining a data-driven and a theory-based approach to generate culture-dependent behaviours for virtual characters. Advances in Culturally-Aware Intelligent Systems and in Cross-Cultural Psychological Studies, pp. 111–142. Springer, Berlin (2018)
80. Madsen, M., Gregor, S.: Measuring human-computer trust. In: 11th Australasian Conference on Information Systems, vol. 53, pp. 6–8. Citeseer (2000)
81. Madsen, M., el Kaliouby, R., Goodwin, M., Picard, R.: Technology for just-in-time in-situ learning of facial affect for persons diagnosed with an autism spectrum disorder. In: Proceedings of the 10th International ACM SIGACCESS Conference on Computers and Accessibility, Assets '08, pp. 19–26. ACM, New York, NY, USA (2008). http://doi.acm.org/10.1145/1414471.1414477
82. Marsella, S., Gratch, J.: Computationally modeling human emotion. Commun. ACM **57**(12), 56–67 (2014)
83. Marsella, S.C., Gratch, J., Petta, P.: Computational models of emotion. In: Scherer, K.R., Bänzinger, T., Roesch, E.B. (eds.) Blueprint for Affective Computing (A Sourcebook), pp. 21–41. Oxford University Press, Oxford (2010)
84. Matson, J., Matheis, M., Burns, C., Esposito, G., Venuti, P., Pisula, E., Misiak, A., Kalyva, E., Tsakiris, V., Kamio, Y., Ishitobi, M., Goldin, R.: Examining cross-cultural differences in autism spectrum disorder: a multinational comparison from Greece, Italy, Japan, Poland, and the United States. Eur. Psychiatry **42**, 70–76 (2017). https://doi.org/10.1016/j.eurpsy.2016.10.007
85. Matsumoto, D.: Cultural influences on the perception of emotion. J. Cross Cult. Psychol. **20**(1), 92–105 (1989)
86. Matsumoto, D.: Cultural similarities and differences in display rules. Motiv. Emot. **14**(3), 195–214 (1990)
87. Matsumoto, D., Yoo, S.H., Nakagawa, S.: Culture, emotion regulation, and adjustment. J. Pers. Soc. Psychol. **94**(6), 925 (2008)
88. Mayer, R.C., Davis, J.H., Schoorman, F.D.: An integrative model of organizational trust. Acad. Manag. Rev. **20**(3), 709–734 (1995)
89. Mazefsky, C.A., Herrington, J., Siegel, M., Scarpa, A., Maddox, B.B., Scahill, L., White, S.W.: The role of emotion regulation in autism spectrum disorder. J. Am. Acad. Child Adolesc. Psychiatry **52**(7), 679–688 (2013)
90. McQuiggan, S.W., Lester, J.C.: Modeling and evaluating empathy in embodied companion agents. Int. J. Hum Comput Stud. **65**(4), 348–360 (2007)
91. Misu, T., Kawahara, T.: Bayes risk-based dialogue management for document retrieval system with speech interface. Speech Commun. **52**(1), 61–71 (2010)
92. Moser, U., Zeppelin, I.v.: Die Entwicklung des Affektsystems. Psyche - Zeitschrift für Psychoanalyse und ihre Anwendungen **50**(1), 32–84 (1996)
93. Nakanishi, J., Sumioka, H., Shiomi, M., Nakamichi, D., Sakai, K., Ishiguro, H.: Huggable communication medium encourages listening to others. In: Proceedings of the Second International Conference on Human-agent Interaction, pp. 249–252. ACM (2014)
94. Nesdale, D., Griffith, J., Durkin, K., Maass, A.: Empathy, group norms and children's ethnic attitudes. J. Appl. Dev. Psychol. **26**(6), 623–637 (2005)
95. Nesdale, D., Milliner, E., Duffy, A., Griffiths, J.A.: Group membership, group norms, empathy, and young children's intentions to aggress. Aggress. Behav.: Off. J. Int. Soc. Res. Aggress. **35**(3), 244–258 (2009)
96. Noddings, N.: Caring: A Relational Approach to Ethics and Moral Education. University of California Press, California (2013)
97. Ogden, B., Dautenhahn, K., Stribling, P.: Interactional structure applied to the identification and generation of visual interactive behavior: robots that (usually) follow the rules. In: International Gesture Workshop, pp. 254–268. Springer (2001)

98. Otaki, H., Otake, M.: Interactive robotic system assisting image based dialogue for the purpose of cognitive training of older adults. In: 2017 AAAI Spring Symposium Series (2017)
99. Ozge Nilay, Y.: Evaluating empathy in artificial agents. In: 8th International Conference on Affective Computing and Intelligent Interaction (ACII) (2019)
100. Paiva, A., Dias, J., Sobral, D., Aylett, R., Sobreperez, P., Woods, S., Zoll, C., Hall, L.: Caring for agents and agents that care: building empathic relations with synthetic agents. In: Autonomous Agents and Multiagent Systems, International Joint Conference on, vol. 2, pp. 194–201. IEEE Computer Society (2004)
101. Paiva, A., Leite, I., Boukricha, H., Wachsmuth, I.: Empathy in virtual agents and robots: a survey. ACM Trans. Interact. Intell. Syst. **7**(3), 11:1–11:40 (2017). http://doi.acm.org/10.1145/2912150
102. Pfeifer, R.: Artificial intelligence models of emotion. Cognitive Perspectives on Emotion and Motivation, pp. 287–320. Springer, Berlin (1988)
103. Picard, R.W.: Future affective technology for autism and emotion communication. Philos. Trans. R. Soc. Lond., B, Biol. Sci. **364**(1535), 3575–3584 (2009)
104. Picard, R.W., Picard, R.: Affective Computing, vol. 252. MIT Press, Cambridge (1997)
105. Plappert, M., Mandery, C., Asfour, T.: Learning a bidirectional mapping between human whole-body motion and natural language using deep recurrent neural networks. Robot. Auton. Syst. **109**, 13–26 (2018)
106. Rao, A.S., Georgeff, M.P.: BDI agents: from theory to practice. ICMAS **95**, 312–319 (1995)
107. Rehm, M.: Developing enculturated agents: pitfalls and strategies. Handbook of Research on Culturally-aware Information Technology: Perspectives and Models, pp. 362–386. IGI Global (2011)
108. Reniers, R.L., Corcoran, R., Drake, R.D., Shryane, N., Völlm, B.A.: The qcae: a questionnaire of cognitive and affective empathy. J. Pers. Assess. **93**, 84–95 (2011)
109. Rieffe, C., Ketelaar, L., Wiefferink, C.H.: Assessing empathy in young children: construction and validation of an empathy questionnaire (emque). Personality Individ. Differ. **49**(5), 362–367 (2010)
110. Robinette, P., Li, W., Allen, R., Howard, A.M., Wagner, A.R.: Overtrust of robots in emergency evacuation scenarios. In: The Eleventh ACM/IEEE International Conference on Human Robot Interaction, pp. 101–108. IEEE Press (2016)
111. Rodríguez, L.F., Ramos, F.: Development of computational models of emotions for autonomous agents: a review. Cogn. Comput. **6**(3), 351–375 (2014)
112. Rutherford, M.D., Baron-Cohen, S., Wheelwright, S.: Reading the mind in the voice: a study with normal adults and adults with asperger syndrome and high functioning autism. J. Autism Dev. Disord. **32**(3), 189–194 (2002)
113. Sanders, T., Oleson, K.E., Billings, D.R., Chen, J.Y., Hancock, P.A.: A model of human-robot trust: Theoretical model development. In: Proceedings of the human factors and ergonomics society annual meeting, vol. 55, pp. 1432–1436. SAGE Publications, Los Angeles (2011)
114. Schaefer, K.: The perception and measurement of human-robot trust (2013)
115. Schwartz, S.: A theory of cultural value orientations: explication and applications. Comp. Sociol. **5**(2–3), 137–182 (2006)
116. Schwartz, S.H.: A theory of cultural values and some implications for work. Appl. Psychol. **48**(1), 23–47 (1999)
117. Shitaoka, K., Tokuhisa, R., Yoshimura, T., Hoshino, H., Watanabe, N.: Active listening system for dialogue robot. JSAI SIG-SLUD Technical Report **58**, 61–66 (2010)
118. Shrestha, M.C., Tsuburaya, Y., Onishi, T., Kobayashi, A., Kono, R., Kamezaki, M., Sugano, S.: A preliminary study of a control framework for forearm contact during robot navigation. In: 2018 27th IEEE International Symposium on Robot and Human Interactive Communication (RO-MAN), pp. 410–415. IEEE (2018)
119. Siau, K., Wang, W.: Building trust in artificial intelligence, machine learning, and robotics. Cut. Bus. Technol. J. **31**(2), 47–53 (2018)
120. Slote, M.: The Ethics of Care and Empathy. Routledge, England (2007)

121. Spiekermann, S.: Digitale Ethik: Ein Wertesystem für das 21. Jahrhundert, Droemer eBook (2019)
122. Spreng, R.N., McKinnon, M.C., Mar, R.A., Levine, B.: The toronto empathy questionnaire: scale development and initial validation of a factor-analytic solution to multiple empathy measures. J. Pers. Assess. **91**(1), 62–71 (2009)
123. Stuck, R.E.: Understanding dimensions of trust between older adults and human or robot care providers. Ph.D. thesis, Georgia Institute of Technology (2017)
124. Takahashi, H., Ban, M., Osawa, H., Nakanishi, J., Sumioka, H., Ishiguro, H.: Huggable communication medium maintains level of trust during conversation game. Front. Psychol. **8**, 1862 (2017)
125. Takano, W., Nakamura, Y.: Statistical mutual conversion between whole body motion primitives and linguistic sentences for human motions. Int. J. Robot. Res. **34**(10), 1314–1328 (2015)
126. Tanaka, H., Adachi, H., Ukita, N., Ikeda, M., Kazui, H., Kudo, T., Nakamura, S.: Detecting dementia through interactive computer avatars. IEEE J. Transl. Eng. Health Med. **5**, 1–11 (2017)
127. Tanaka, H., Negoro, H., Iwasaka, H., Nakamura, S.: Embodied conversational agents for multimodal automated social skills training in people with autism spectrum disorders. PLOS ONE **12**(8), 1–15 (2017). https://doi.org/10.1371/journal.pone.0182151
128. Tanveer, M.I., Lin, E., Hoque, M.E.: Rhema: A real-time in-situ intelligent interface to help people with public speaking. In: Proceedings of the 20th International Conference on Intelligent User Interfaces, IUI '15, pp. 286–295. ACM (2015). http://doi.acm.org/10.1145/2678025.2701386
129. Tapus, A., Matarić, M.J.: Emulating empathy in socially assistive robotics. In: AAAI Spring Symposium: Multidisciplinary Collaboration for Socially Assistive Robotics, pp. 93–96 (2007)
130. Thill, S., Pop, C.A., Belpaeme, T., Ziemke, T., Vanderborght, B.: Robot-assisted therapy for autism spectrum disorders with (partially) autonomous control: challenges and outlook. Paladyn **3**(4), 209–217 (2012)
131. Trommsdorff, G., Friedlmeier, W., Mayer, B.: Sympathy, distress, and prosocial behavior of preschool children in four cultures. Int. J. Behav. Dev. **31**(3), 284–293 (2007). https://doi.org/10.1177/0165025407076441
132. Ullman, D., Malle, B.F.: Human-robot trust: just a button press away. In: Proceedings of the Companion of the 2017 ACM/IEEE International Conference on Human-Robot Interaction, pp. 309–310. ACM (2017)
133. Van Mulken, S., André, E., Müller, J.: The persona effect: how substantial is it? People and Computers XIII: Proceedings of HCI, vol. 98, pp. 53–66 (1998)
134. Vargas, P.A., Fernaeus, Y., Lim, M.Y., Enz, S., Ho, W.C., Jacobsson, M., Aylett, R.: Advocating an ethical memory model for artificial companions from a human-centred perspective. AI Soc. **26**(4), 329–337 (2011)
135. B M de Waal, F., D Preston, S.: Mammalian empathy: behavioural manifestations and neural basis. Nat. Rev. Neurosci. **18** (2017). https://doi.org/10.1038/nrn.2017.72
136. Watson, D., Clark, L.A., Tellegen, A.: Development and validation of brief measures of positive and negative affect: the PANAS scales. J. Pers. Soc. Psychol. **54**(6), 1063–1070 (1988)
137. Wey Smola, K., Sutton, C.D.: Generational differences: revisiting generational work values for the new millennium. J. Organ. Behav.: Int. J. Ind., Occup. Organ. Psychol. Behav. **23**(4), 363–382 (2002)
138. Wilks, Y. (ed.): Close Engagements with Artificial Companions: Key Social, Psychological, Ethical and Design Issues, vol. 8. John Benjamins Publishing, Amsterdam (2010)
139. Yamada, T., Matsunaga, H., Ogata, T.: Paired recurrent autoencoders for bidirectional translation between robot actions and linguistic descriptions. IEEE Robot. Autom. Lett. **3**(4), 3441–3448 (2018)

140. Yoshino, K., Kawahara, T.: Conversational system for information navigation based on pomdp with user focus tracking. Comput. Speech Lang. **34**(1), 275–291 (2015)
141. Yoshino, K., Murase, Y., Lubis, N., Sugiyama, K., Tanaka, H., Sakti, S., Takamichi, S., Nakamura, S.: Spoken dialogue robot for watching daily life of elderly people. In: Proceedings of International Workshop on Spoken Dialogue Systems Technology (IWSDS) 2019 (2019)
142. Yoshino, K., Tanaka, H., Sugiyama, K., Kondo, M., Nakamura, S.: Japanese dialogue corpus of information navigation and attentive listening annotated with extended iso-24617-2 dialogue act tags. In: Proceedings of the Eleventh International Conference on Language Resources and Evaluation (LREC-2018) (2018)
143. Young, S., Gašić, M., Keizer, S., Mairesse, F., Schatzmann, J., Thomson, B., Yu, K.: The hidden information state model: a practical framework for pomdp-based spoken dialogue management. Comput. Speech Lang. **24**(2), 150–174 (2010)
144. Zhao, R., Papangelis, A., Cassell, J.: Towards a dyadic computational model of rapport management for human-virtual agent interaction. In: International Conference on Intelligent Virtual Agents, pp. 514–527. Springer (2014)

Chapter 3
Multimodal Machine Learning for Social Interaction with Ageing Individuals

Louis-Philippe Morency, Sakriani Sakti, Björn W. Schuller, and Stefan Ultes

3.1 Introduction

Multimodal machine learning (MMML) is a vibrant multi-disciplinary research field. It addresses some of the original goals of artificial intelligence (AI) by integrating and modelling multiple communicative modalities, including linguistic, acoustic and visual messages amongst other multisensorial information such as physiological signals or symbolic information, e.g., concerning contextual cues. With the goal of better understanding and modelling the behaviours of ageing individuals, this research field brings some unique challenges for multimodal researchers. This comes with the heterogeneity of the potentially mixed symbolic and signal-type data and the contingency often found between modalities. In this chapter, we identify four key challenges necessary to enable multimodal machine learning, in particular, considering ageing individuals:

(1) *multimodal*, this modelling task includes multiple relevant modalities which need to be represented, aligned and fused. Often, these are asynchronous, potentially in dynamic manner and/or operating on different time-scales;

L.-P. Morency
Carnegie Mellon University, Pittsburgh, USA
e-mail: morency@cs.cmu.edu

S. Sakti
Nara Institute of Science and Technology, Ikoma, Japan
e-mail: ssakti@is.naist.jp

B. W. Schuller
Chair EIHW, University of Augsburg, Germany & GLAM, Imperial College London, Augsburg, UK
e-mail: schuller@ieee.org

S. Ultes (✉)
Mercedes-Benz AG, Sindelfingen, Germany
e-mail: stefan.ultes@daimler.com

J. Miehle et al. (eds.), *Multimodal Agents for Ageing and Multicultural Societies*,
https://doi.org/10.1007/978-981-16-3476-5_3

(2) *high variability*, this modelling problem expresses high variability given the many social contexts, large space of actions and possible physical or cognitive impairment;
(3) *sparse and noisy resources*, this modelling challenge addresses unreliable sensory data and the limitation and sparseness of resources that are specific for the special user group of ageing individuals; and
(4) *concept drift*, where two types of drift were identified, namely, on the group level (as the target group of usage is not fully known at the moment of development of according interfaces given that it is yet to age) and the individual level (given that ageing may lead to drifting behaviour and interaction preferences throughout the ongoing ageing effect).

These four challenges come together when we build an evaluation plan that enables, at the same time, a strategy to include the broader machine learning community in this effort. This research agenda will enable more effective and robust modelling technologies as well as development of socially competent and culture-aware embodied conversational agents for elderly care. In the following, we will address these four challenges in the named order one-by-one. Concluding, we will summarize the discussion of these.

3.2 Multimodal

An important aspect of ageing research is to develop foundational methods to analyze, model and represent multimodal information. The unique and challenging aspect of multimodal research is the heterogeneity of the multimodal data and the challenges in integrating and interpreting such heterogeneous data coming from multiple modalities. Creating computers able to understand this multimodal data brings many fundamental problems which can be grouped into five main classes, following the taxonomy of Baltrušaitis et al. [4]: representation, alignment, fusion, translation and co-learning.

Representation: Learning how a computer can represent numerically the heterogeneous data from multiple modalities. These computational representations should be designed for both efficient modelling and better visualization. For example, a joint representation of how a person looks and sounds when they are happy will allow computers to better recognize human emotions. These joint representations will be the most efficient when they can take advantage of the natural dependencies between modalities. Another objective is to improve the interpretability of multimodal data. By identifying commonalities and differences between multimodal data, a multimodal representation provides an avenue to bridge the gap between continuous versus discrete data and numerical versus symbolic data.

Alignment: The process of establishing spatial and/or temporal connections between events across modalities. For example, when reading the caption of an image, alignment is the process where words are linked to specific objects or groups of objects in the image. Other examples of alignment include automatic video capturing and identifying the acoustic source in a video. One challenge in alignment is dealing with data stream with different sampling rates (e.g., continuous signals versus discrete events). Alignment may require defining a similarity metric between modalities to identify the connection points. The alignment can be temporal, as when we align the audio and images of a video, or it can be spatial alignment, as when we try to morph between two face images.

Fusion: The combination of information coming from two or more sources to uncover or predict a pattern or trait of interest. Examples of multimodal fusion include multimodal emotion recognition, audiovisual speech recognition and audiovisual speaker verification. The information can be redundant which helps increase robustness, or complementary which often helps increase accuracy. The challenges are multiple since the modalities do not need to be synchronized or even have the same sampling rate. Some modalities may be incomplete with missing information. Some modalities may provide continuous streams of data, while other modalities may be intrinsically discrete providing information about given events.

Translation: The transformation or mapping from one modality into another. Examples include speech-driven animation, and text-based image retrieval. The foundational methods learn the relationship between streams of data, capturing their dependencies. The goal of translation can be generative in nature, creating a new instance in one modality given information from another modality. It can also be descriptive models, where one modality is used to increase the characterization of another modality (e.g., describing the information conveyed in an image).

Co-learning: The transfer of knowledge from one modality to help with the prediction or modelling task in another modality. Co-learning examples are more technical in nature. One of the most popular examples these days is the use of language to help generalization of computer vision algorithms, specifically for object recognition. Co-learning aims to leverage the rich information in one modality, in the learning of another modality, which may have only limited resources (e.g., small number of examples with limited annotations or noisy input). Example of co-learning algorithms include co-training, zero-shot learning and concept learning.

Problems on multimodal processing may involve combination of these categories. Importantly, crosscutting research in these areas will open opportunities to better understand and interpret multimodal data across domains, serving as instrumental tool for the community. These tools can be generic, working across problems. They can also be specific to determined problems or modalities.

3.3 High Variability

Realizing multimodal machine learning in highly variable environments adds to the challenges described previously. Variations may impact the performance of machine learning models and may be categorized by

Variation on the input Deriving suitable knowledge from observed situations itself is already quite challenging as relevant information needs to be separated from irrelevant and redundant information. In a highly variable environment, this task is even more difficult, e.g., as some pieces of information is only relevant for a small amount of cases but irrelevant for the rest. Another aspect is the increased size that is necessary to model the input in a highly variable environment.

Variation on the output For classification or decision making tasks, the number of labels or actions is directly linked to the difficulty of the learning problem. With each added variation on theses sets, the number of classes or actions increases even further and thus poses a serious challenge on the learning task as these become more fine-grained or may even overlap in some cases. Thus, it is more difficult to discriminate between them.

Both types of high variability poses crucial challenges towards realizing multimodal machine learning approaches, especially within the context of social interaction with the focus group of ageing individuals. Some factors that cause this high variability are:

Many social contexts Elderly citizens engage in many social contexts—just as the non-elderly. They do volunteering, or are active members in local clubs in addition to regular participation in social activities like meeting with friends and family. Additionally, their living conditions might at the same time include elderly care homes or other care-taking facilities resulting in a broad range of social contexts.

Physical or cognitive impairment The existence of physical or cognitive disabilities becomes more prevalent with increasing age: "Seniors are almost twice as likely to have a disability as those of working age" [13, 14]. This comes along with a unique nature of communication on both sides of the communication channel: input to a social agent may come in many different ways and social agents must use communication means that are adequate for the given situation. Thus, a social agent needs to offer a wide range of options for communication.

Reluctance to adapt Seniors are often reluctant to adapt to new situations and exhibit stubbornness due to many reasons, e.g., feeling of losing their independence, fear of losing control of their lives, feeling depressed about the deaths of spouse, friends and/or family, feeling of being left out of the family, or fear of their own mortality [1]. Social agents must thus be able to provide interaction tailored to the specific needs of many different concrete users.

While high variability is not unique to multimodal machine learning setups, the additional modalities further increase the variability and extend it to multiple communication channels, thus adding to the challenges described in Sect. 3.2.

The high variability has a very simple yet very problematic effect on machine learning models: they either are less likely to generalize well or need an increasing amount of diverse and adequate data that contains all relevant situations and information. However, a simple increase in data does not necessarily lead to well-working models. The important part here is that the factors that cause the variability are part of the input feature set of the machine learning models. While approaches that learn to extract the relevant information like deep learning model these factors implicitly, other approaches create the need for explicit modelling.

For applications that make use of multimodal machine learning to derive knowledge from observed situations, this either results in a huge input space or in separate models, either one for each category of variation[1] or arranged hierarchically. More individualized models are required that do not follow the "one-fits-all" paradigm but are able to grasp all the relevant information on the input side, e.g., [2, 23].

Artificial agents that interact with seniors additionally are faced with the challenge of exhibiting adequate behaviour for the given context. Hence, they need to understand the social context and they need to be able to react accordingly, e.g., the non-elderly might be more forgiving if the system behaviour is socially awkward.

3.4 Sparse and Noisy Resources

In multimodal machine learning, each modality may provide complementary information and cross-modality feature learning could help the model to understand more. However, to extract generalizable features in a supervised learning approach and ensure a robust perception of the overall information, multimodal machine learning requires a large-scale labelled training dataset. Unfortunately, such data is often unavailable, especially for multimodal resources needed to construct machine learning for social interaction with ageing individuals.

Several issues make data collection in ageing individuals become critical bottlenecks.

- Although the population of ageing individuals is growing in all regions of the world, it is still minimal compared to the general population. This makes it hard to collect sufficient amounts of their social interaction in real life.
- Cognitive or physical deficits can lead to the inability to perform some assessments, which leads to incomplete or partially-observed data.
- Imperfect sensory data are often unavoidable in real-world environments, resulting in unreliable measures and noisy data.
- Data preprocessing and manual labelling or annotation are expensive and time-consuming. If we cannot bear the cost, a large amount of data will remain unlabelled.

Consequently, to handle such sparse and noisy data, other approaches beyond traditional supervised learning fashion become necessary. Extensive researches are currently focused on a learning algorithm that can be performed without the need for expensive supervision. To date, several approaches have been proposed, including transfer learning, semi-supervised learning, self-supervised learning and active learning.

---

[1] A requirement for separate models is that the observed variability can be divided into distinct categories.

Transfer learning The study of transfer learning is motivated by the fact that humans can intelligently apply their knowledge learned from previous problems to solve new tasks faster or with better solutions. In transfer learning, the knowledge or information of an already trained machine learning model with a sufficient labelled training dataset is reused or applied in a new task with limited resources. Therefore, instead of constructing the model from scratch using a minimal dataset, we begin with patterns learned from solving a related task. Palaskar et al. applied transfer learning for audiovisual scene-aware dialogue [16]. Specifically, they developed a hierarchical attention framework to fuse contributions from different modalities and utilized the framework to generate textual summaries from multimodal sources (i.e., videos with accompanying commentary). Wolf et al. also introduced a transfer learning approach to generative data-driven dialogue in conversational agents [26]. The framework was called TransferTransfo, which is a combination of a Transfer learning scheme and a high-capacity Transformer model.

Semi-supervised learning A semi-supervised learning model aims to make effective use of all of the available data, not just the labelled data, but also unlabelled data. The primary method is to train a model with the labelled data, then let the model label the unlabelled examples. Then, finally, retrain the model with the additional training dataset produced by the model. In a multimodal framework, Effendi et al. proposed semi-supervised learning for cross-modal data augmentation via a multimodal chain, and addressed the problems of speech-to-text, text-to-speech, text-to-image and image-to-text [7]. A study by Tseng et al. showed that semi-supervised techniques could reduce the need for intermediate-level annotations in training neural task-oriented dialogue models [21].

Self-supervised learning Self-supervised learning provides a viable solution when labelled training data is scarce. The framework is commonly performed to learn the relations or correlations between inputs, such as predicting word context or image rotation, for which the target can be computed without supervision. In spoken dialogue research, Wu et al. utilized self-supervised learning for inconsistent dialogue order detection by explicitly capturing the conversation's flow in dialogues [27]. A study by Li et al. also performed a self-supervised method for a multimodal dialogue generation model [12]. Specifically, they adopted the multitask learning, including response language modelling, video-audio sequence modelling and caption language modelling, to learn joint representations and generate informative and fluent responses.

Active learning Active learning is one way to address the problem of a limited annotation budget by actively enlarge the training datasets. The basic idea is to ease the data collection process by automatically deciding which instances an annotator should label to train an algorithm quickly and effectively. In the context of dialogue research, Hiraoka et al. proposed active learning in creating dialogue examples [10]. Gasic and Young [9] used active learning to speed reinforcement learning of the dialog system policy. A study by Rudovic et al. also offered a multimodal active learning approach, in which, deep reinforcement learning is used to find an optimal policy for active selection of the user's data [19].

3.5 Concept Drift

In a longitudinal human-machine interaction setting, one likely changes his/her characterizing attributes [24], preferences [5], behaviours and how these are realized and portrayed. In particular, when ageing, elderly may be gradually affected by increasing mental and physical limitations or confronted with abrupt life changes in life circumstances such as by moving to care-taking facilities. This renders them a user group of likely occurrence of concept drift in continued every-day multimodal social interaction scenarios.

In machine learning, concept drift describes a non-stationary learning problem over time. Hence, it primarily refers to the learning targets the model learns to classify or predict and their change over time [22, 25]. In a classification task, this could mean that the classes and their relation to the 'input' change. Similarly, in a regression problem, the numerical target would 'drift' in the sense of change over time. In particular, this becomes challenging, if this change is unknown and it has to be detected in the first place. In any case, such a drift demands for adaptation of the learnt models over time. This could, e.g., be realized in a life-long learning process [17] or by some suited means of incremental transfer learning to adapt to the changing target based on the knowledge previously required to best classify the former 'old' target. In a broader definition, concept drift could, however, also refer to input drift, such as when the input type changes over time. In addition, as outlined, one also potentially needs to detect the presence of concept drift in the first place [11, 22].

Coming back to dealing with (multimodal) interaction, an (elderly) user may as mentioned alter her behaviour over time, hence, also enforcing a model change in time (the concept drift) [8]. For our particular task of interest—multimodal machine learning for social interaction with ageing users—we see two main types of drift:

(1) drift on the group level. This is understood in the sense of the elderly as population or sub-groups thereof, such as clustered by gender, age groups, etc., as the target group including its characterizing attributes, preferences and behaviour patterns of interaction is not fully known at the moment of development of according interfaces given that it is yet to age, and
(2) on the individual level. The motivation for this level is given by the fact that ageing as a process may lead to drifting behaviour and interaction preferences throughout the ongoing individual ageing, potentially driven, e.g., by the named potential limitations in cognitive and/or physical abilities. Note that change of preferences and, in particular, change in (elderly) user behaviour can be both gradual or abrupt, given potential sudden occurrence of named potential limitations such as given by a gradual progressing age-induced disorder or following a fall or stroke or a change in living conditions. These two kinds of concept drift—abrupt or gradual—are generally known as the two types of occurrences of concept drift [22].

While the problem of concept drift has practically not been investigated in the field of social interaction with a human–computer interface or agent even for any user group, the methods available to deal with the problem start to find interest in this and the broader application domain. As an example, life-long learning has been considered for emotion recognition [17]. Further, for recognition of behaviour of elderly in smart homes, the authors in [28] suggest to integrate activity duration into the learning process learning to cope with concept drift. More broadly looking into human behaviour modelling in general, the author in [15] names active learning an interesting option, given that incremental learning is often chosen as a mean to handle concept drift. This could include cooperation with the elderly individuals to label new most informative data points with their aid during usage.

However, in other domains of multimodal machine learning, the problem has been handled more frequently, such as in [18]. The authors see the problem of time-variant input/output relation as dynamic optimisation problem (DOP). They compare two methods that both improve robustness in the presence of concept drift: a time-window solution to train exclusively on most recent data, and a "time-as-covariate" approach modelling time as additional input variable. Considering a number of benchmark functions, they find the former better suited in case of higher-dimensional multimodal cases, and the later in the opposite case of lower dimensional multimodal tasks, as those make it easier to co-model time as input.

Further, in the general machine learning literature, one finds manifold solutions, such as in [6], where the authors suggest semi-supervised learning as a mean to handle concept drift. In addition, instance selection, instance weighting and ensemble learning [22] are popular 'traditional' methods for the detection and adaptation [3]. In particular for the recently popular deep learning approaches, online learning methods have been proposed such as hedge back-propagation [20]. For further insights, a number of overviews exist such as [29].

As a conclusion, concept drift will be a real-world challenge for multimodal social interaction over pro-longed usage—likely in particular for an ageing elderly user group. Methods to cope with the problem do exist, but yet have to be integrated and evaluated more in this application context.

3.6 Conclusion

Multimodal machine learning itself already contains hard challenges in general. These have been described in this chapter and have been embedded into the context of social interaction with ageing individuals. On top of *representation*, *alignment*, *fusion*, *translation* and *co-learning*, we identified *high variability*, *sparse and noisy resources* and *concept drift* as the most important topics. *High variability* is described as variation on the input or the output leading to an increase in complexity of the machine learning model. *Sparse and noisy resources* further require advanced techniques like transfer learning or active learning. Especially the latter is also proposed to alleviate some of the problems arising from *concept drift* where important properties of the problem change over time, and thus require special handling on the modelling side.

References

1. Assisted senior living: dealing with stubbornness. https://www.assistedseniorliving.net/caregiving/dealing-with-stubbornness/. Accessed 02 Oct 2019
2. AVEC '19: Proceedings of the 9th International on Audio/Visual Emotion Challenge and Workshop. Association for Computing Machinery, New York, NY, USA (2019)
3. de Barros, R.S.M., de Carvalho Santos, S.G.T.: An overview and comprehensive comparison of ensembles for concept drift. Inf. Fusion **52**, 213–244 (2019)
4. Baltrušaitis, T., Ahuja, C., Morency, L.P.: Multimodal machine learning: a survey and taxonomy. IEEE Trans. Pattern Anal. Mach. Intell. **412**, 423–443 (2018)
5. Campigotto, P., Passerini, A., Battiti, R.: Handling concept drift in preference learning for interactive decision making. HaCDAIS **2010**, 29 (2010)
6. Dyer, K.B., Polikar, R.: Semi-supervised learning in initially labeled non-stationary environments with gradual drift. In: The 2012 International Joint Conference on Neural Networks (IJCNN), pp. 1–9. IEEE (2012)
7. Effendi, J., Tjandra, A., Sakti, S., Nakamura, S.: Listening while speaking and visualizing: improving asr through multimodal chain. In: 2019 IEEE Automatic Speech Recognition and Understanding Workshop (ASRU) pp. 471–478 (2019)
8. Esposito, F., Basile, T.M., Di Mauro, N., Ferilli, S.: Machine learning enhancing adaptivity of multimodal mobile systems. In: Multimodal Human Computer Interaction and Pervasive Services, pp. 121–138. IGI Global (2009)
9. Gašić, M., Young, S.: Gaussian processes for POMDP-based dialogue manager optimization. IEEE/ACM Trans. Audio, Speech, Lang. Process. **22**(1), 28–40 (2014)
10. Hiraoka, T., Neubig, G., Yoshino, K., Toda, T., Nakamura, S.: Active Learning for Example-Based Dialog Systems, chap. Dialogues with Social Robots: Enablements, Analyses, and Evaluation, pp. 67–78 (2017)
11. Klinkenberg, R., Joachims, T.: Detecting concept drift with support vector machines. In: ICML, pp. 487–494 (2000)
12. Li, Z., Li, Z., Zhang, J., Feng, Y., Niu, C., Zhou, J.: Bridging text and video: a universal multimodal transformer for video-audio scene-aware dialog. AAAI2020 DSTC8 workshop (2020)
13. Morris, S., Fawcett, G., Brisebois, L., Hughes, J.: Canadian survey on disability reports: a demographic, employment and income profile of Canadians with disabilities aged 15 years and over (2017). https://www150.statcan.gc.ca/n1/pub/89-654-x/89-654-x2018002-eng.htm. Accessed 27 May 2020
14. Murman, D.L.: The impact of age on cognition. Semin. Hear. **36**(03), 111–121 (2015). https://doi.org/10.1055/s-0035-1555115
15. Padmalatha, E., Reddy, C., Rani, P.: Mining concept drift from data streams by unsupervised learning. Int. J. Comput. Appl. **117**(15) (2015)
16. Palaskar, S., Sanabria, R., Metze, F.: Transfer learning for multimodal dialog. Comput. Speech Lang. **64**, 101093 (2020). https://doi.org/10.1016/j.csl.2020.101093
17. Ren, Z., Han, J., Cummins, N., Schuller, B.: Enhancing transferability of black-box adversarial attacks via lifelong learning for speech emotion recognition models. In: Proceedings INTERSPEECH 2020, p. 5. ISCA, ISCA, Shanghai, China (2020)
18. Richter, J., Shi, J., Chen, J.J., Rahnenführer, J., Lang, M.: Model-based optimization with concept drifts. In: Proceedings of the 2020 Genetic and Evolutionary Computation Conference, pp. 877–885 (2020)
19. Rudovic, O., Zhang, M., Schuller, B., Picard, R.W.: Multi-modal active learning from human data: a deep reinforcement learning approach (2019). https://arxiv.org/abs/1906.03098
20. Sahoo, D., Pham, Q., Lu, J., Hoi, S.C.: Online deep learning: learning deep neural networks on the fly (2017). arXiv:1711.03705
21. Tseng, B.H., Rei, M., Budzianowski, P., Turner, R., Byrne, B., Korhonen, A.: Semi-supervised bootstrapping of dialogue state trackers for task-oriented modelling. In: Proceedings of the 2019

Conference on Empirical Methods in Natural Language Processing and the 9th International Joint Conference on Natural Language Processing (EMNLP-IJCNLP), pp. 1273–1278. Hong Kong, China (2019)
22. Tsymbal, A.: The problem of concept drift: definitions and related work. Computer Science Department, Trinity College Dublin **106**(2), 58 (2004)
23. Wagner, J., Lingenfelser, F., Baur, T., Damian, I., Kistler, F., André, E.: The social signal interpretation (ssi) framework: multimodal signal processing and recognition in real-time. In: Proceedings of the 21st ACM International Conference on Multimedia, pp. 831–834. ACM (2013)
24. Webb, G.I., Pazzani, M.J., Billsus, D.: Machine learning for user modeling. User Model. User-Adapt. Interact. **11**(1–2), 19–29 (2001)
25. Widmer, G., Kubat, M.: Learning in the presence of concept drift and hidden contexts. Mach. Learn. **23**(1), 69–101 (1996)
26. Wolf, T., Sanh, V., Chaumond, J., Delangue, C.: Transfertransfo: a transfer learning approach for neural network based conversational agents. NeurIPS 2018 CAI Workshop (2019)
27. Wu, J., Wang, X., Wang, W.Y.: Self-supervised dialogue learning. In: Proceedings of the 57th Annual Meeting of the Association for Computational Linguistics (ACL), pp. 3857–3867. Florence, Italy (2019)
28. Zhang, S., McClean, S., Scotney, B., Chaurasia, P., Nugent, C.: Using duration to learn activities of daily living in a smart home environment. In: 2010 4th International Conference on Pervasive Computing Technologies for Healthcare, pp. 1–8. IEEE (2010)
29. Žliobaitė, I.: Learning under concept drift: an overview (2010). arXiv:1010.4784

Chapter 4
Multimodal and Multicultural Field Agents: Considerations for "outside-the-lab" Studies

Matthias Rehm

4.1 Introduction

One of the application scenarios for multimodal and multicultural agents that is discussed in this book is support in care settings, both at home and in institutionalized care. This field presents a complex environment with many practices and routines that are specific to the given context and might have a strong influence on the use and adoption of agent technology. It also presents a context where field research is necessary during several steps of the development process of an agent system: to gain contextual knowledge about, to define the problem space, to allow for a user-centered design of system and services and last but not least to test and evaluate the agent under realistic conditions. A positive example is presented by Voit and colleagues [31], who were interested in assessing the effect the experimental method has on the user's assessment of a prototype and compared among others, labs studies with in-situ experiments. They found that there are significant effects of the study method and that prototypes tended to be rated better in in-situ settings.

Most user studies with agents on the other hand are done in the lab, i.e., in a decontextualized, controlled setting, often with participants that do not originate from the intended target group, making, e.g., use of students as convenience samples. In our experience, this might lead to problems, when systems are deployed in the actual context of their use. In one of our projects, e.g., we developed personal robots for citizens with brain damage [26]. It became clear very fast during workshops with the target users and their care personnel that we cannot assume a dyadic interaction between the robot and the user. Instead, we had to model the interaction behavior of the robot with the citizen and his/her carer as a team in mind.

Although this is an extreme example of development and testing outside the lab, it shows an advantage of field studies in general, i.e., their higher ecological validity.

M. Rehm (✉)
Aalborg University, Aalborg, Denmark
e-mail: matthias@create.aau.dk

J. Miehle et al. (eds.), *Multimodal Agents for Ageing and Multicultural Societies*,
https://doi.org/10.1007/978-981-16-3476-5_4

Lab studies on the other hand have the advantage of a smaller and more precise focus on a limited number of variables and control over the design, procedure, and noise variables, which is an indisputable advantage for developing and testing the technical aspects of a system.

The need for field studies arises, when we are actually interested in evaluating an agent system in the context of use for which it is envisioned. Such a field deployment will generate knowledge for the further development and often it will also generate new research questions [27]. The advantage is obviously that we will be able to explore the interactions between the technology that has been developed and the context in which it is envisioned to be deployed, including elements like the social practices in which it will be embedded (or which it will in the worst case disrupt beyond repair), additional users with unforeseen roles (e.g., as helper, bystander, commentator, etc.), or physical conditions like too much light to see the screen, too much noise to hear the audio output, no stable wifi connection, and may others. Such in-situ evaluations allows for studying the real impact on the target population and on the environment. They also allow for long-term studies of acceptance, adoption, and appropriation of the developed technology. A crucial requirement though is a robust and useful prototype, as participants should not be frustrated by frequent breakdowns or bored to death by too simple interaction routines.

From this enumeration, it is already clear that we will seldom be able to conduct a controlled experiment in the field. Instead, we will need a toolbox of methods that allows for assessing the technical aspects we are interested in, while at the same time ensuring to capture unforeseen information for later analysis and adaptation of the system. Running a successful field study also requires having a robust working prototype available, both in terms of user interaction but also durability. Additionally, field studies are expensive because they are time-consuming and resource intensive and require the target population to actively engage with the system over the whole period of the field study.

While there is increasing focus on studies outside the lab in the HCI community (e.g., [10, 31, 32]) reporting on studies but also starting to discuss methods for in the wild studies, we see a lack of similar work in the agent community. A review combining the data from Norouzi and colleagues [19] concerning user studies presented at the International Conference on Intelligent Virtual Agents for 2001–2015 with the date from the last four years (2016–2019) reveals that in these 19 years only 26 out of 361 user studies were conducted as field studies (ca. 7%, Fig. 4.1). Figure 4.2 shows the development of number of field studies over the years, and highlights that this number is stable, ranging from 0 to 4 with not discernible trend. We can only speculate about the reasons for this and have three informed guesses. The first one is that field studies require robustly working prototypes that realize the whole processing pipeline from user input over behavior selection to multimodal output. Often studies are concerned only with certain aspects of this pipeline, and thus it makes sense to conduct a lab study instead. The second guess has to do with the effort that is needed to conduct field studies both in terms of time and costs. And last but not least, there might be an uncertainty about suitable methods and best practices in conducting field studies. The rest of this chapter is concerned with this last aspect.

Fig. 4.1 Number of field studies compared to total number of user studies, IVA conference 2001–2019

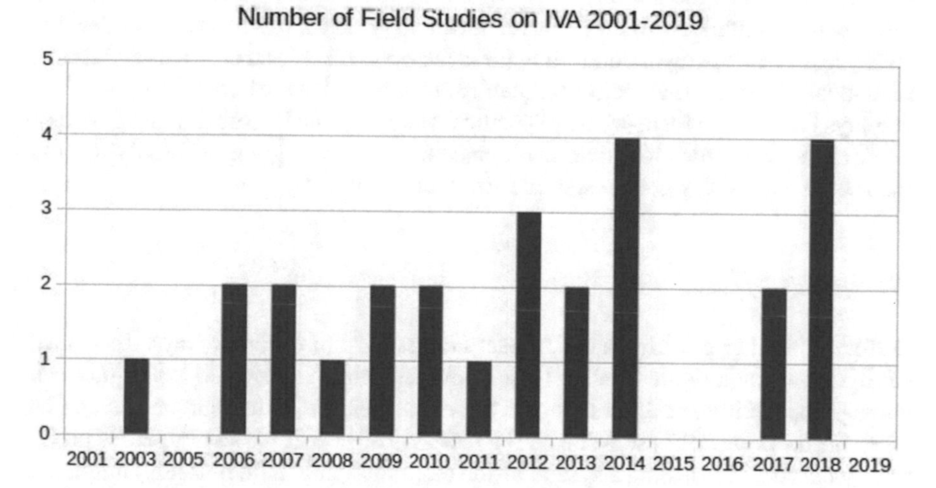

Fig. 4.2 Number of field studies presented at the IVA conference 2001–2019

In order to lay the ground for more field deployments of agents (virtual or robotic), this chapter will start with presenting some relevant work from HCI concering studies outside the lab in the next section. The rest of this chapter then investigates methods for data collection in the field, as well as ethical aspects of field studies, especially in care contexts.

4.2 Lessons to Be Learned from HCI Field Studies

We have to start with clarifying some notions used for research that is done outside the lab. In the HCI community, we can find the notion of studies conducted "in the wild". While this has been in use for some time, the notion has been critically discussed recently as inappropriate, especially in relation to studies with indigenous groups [28]. Thus, we work here with the more neutral terms of field studies or field deployment, where the first denotes a dedicated short term evaluation in the field, and the last signifies a long-term study, where the system is left in the field for extended use.

In this section, we discuss some insights from previous work in HCI, mainly in the are of public displays, that are relevant for designing and conducting field studies.

4.2.1 External Factors

Moving outside of the lab increases the chance that external factors can influence the way users interact with the deployed technology. Thus, it is necessary to develop strategies of addressing or controlling these factors. Mäkelä and colleagues present a meta-analysis of external factors frequently encountered in the deployment of public displays [10]. While focusing on a specific technology, the factors described in their work are relevant for most field deployments and we are going to briefly discuss them here, sometimes giving examples from our own work.

Weather

Different weather conditions can impact field studies in different ways. Too much sunlight can make it impossible to see anything on a screen, too much rain can damage the equipment. Depending on the exact location, either sun or rain can be effecting the number of users that can be recruited on site. If it's sunny, people prefer to stay outside, thus testing a system inside (e.g., shopping mall, museum) might not attract enough participants. If it's rainy, people prefer to be inside, thus an outside test will face difficulties attracting participants. We experienced this challenge during a field study for an agent-based treasure hunt at an art museum [23]. The system was geared towards families, which meant we were able to recruit the most participants on weekends when it was rainy weather resulting in difficult scheduling processes for the study, where researchers, museum staff, and the weather had to be coordinated. The weather is, of course, a problem for field studies that rely on sophisticated sensor technologies like computer vision, where, e.g., too much sunlight can also be detrimental to the accuracy of the system. We experienced the Own experience (interactive lamp: due to long light in summer, test only possible on weekends (people staying out long enough + in a good mood); agent-based treasure hunt at the museum REF: best possibilities on rainy days, few families on sunny days).

Events

Sometimes events happen during field studies, e.g., festivals, workshops, or parties. Such events have two direct effects. The change the demographics in the area as they attract specific crowds to the space in which the deployment takes place. Also, they increase the flow of people, and thus the possibility to test with more users. But they also pose a demand on the prototype to be able to handle group interactions and pose a potential threat for the security of the equipment (see also vandalism below). We already mentioned the challenges of weather when we discussed our museum prototype. We were also able to exploit school outings to museum, which happened on a regular basis during the week, allowing us to recruit additional participants, but with a different social demographics then with the original envisioned family outings.

Surroundings

The build environment around the deployment space can influence the field study. This includes buildings, traffic, vegetation, etc. Contrary to events, these are permanent and have to be taken into account when designing the deployment. Examples of effects include a lack of electricity or connectivity, reflections from windows counteracting vision systems, or diesel fumes clogging filters in equipment. We experienced a concrete example when we were working on an educational robotic system with neurocenter and residency for adolescents with congenital brain damage [11]. The system needed internet connectivity to work properly, which was available in principle. But the institution's IT service was handled by the municipality, which was not willing to give access to the robots resulting in the need of setting up a separate private network to handle communication for the system.

Space

The physical structure of the immediate deployment area has an impact on the flow of users and their willingness to interact (see also the notion of comfort spaces by Fischer and Hornecker [8]). Examples of problematic spaces mentioned in [10] are, e.g., limited visibility due to placement in corners obstructed from view or placement next to an elevator, where people started interacting with the system while waiting and then abruptly left when the elevator arrived resulting in a number of interrupted trials.

Inhabitants

Apart from participants for the field study, people frequent the deployment area on a regular basis, e.g., janitors, cleaning staff, residents, etc. If they are not properly informed about the deployment and its goals, this can lead to unwanted interference, e.g., shutting down systems to save energy or blocking interaction spaces due to other routines. This can pose a threat during unattended long-term field deployments.

Vandalism

If equipment is used in public areas, destruction of equipment can happen. This is difficult to prevent, especially in unattended field deployments. Other types of vandalism concern intended wrong usage of system, e.g., by uploading rude content in a system that allows for distributing user generated content. A moderator might be necessary in such a case.

Mäkelä and colleagues [10] identify four strategies for addressing the challenges posed by these external factors.

1. Ignoring: Nothing is done to address the observed effect. The assumption for employing this strategy is most likely that it will only happens once.
2. Adapting: Some property of the deployment is changed temporarily in order to meet the external factor. This might lead to a sub-optimal system performance.
3. Solving: To solve the problem with an external factor, this can either be removed, or a property of the deployment can be changed permanently.
4. Embracing: The external factor can also be taken advantage of, thus turning it into a positive aspect of the deployment.

In their review, they can see that the first two strategies are most frequently adapted, i.e., 78% of cases use ignoring or adapting as coping strategies.

4.2.2 Experimenter's Influence

During field studies, the researchers are at least for some time part of the deployment as an observer, recruiter, or trouble solver. Williamson and Williamson [32] analyze the different roles of an experimenter and their effects on field deployments. The roles defined for the experimenter have direct consequences for the type of data that can be collected and behavioral biases of participants. They distinguish three possible roles during a public evaluation.

Steward Observer

A steward observer takes on an active role in recruiting and guiding participants allowing a certain level of control over the experience. Moreover, it becomes possible to collect additional data, e.g., through interviews and questionnaires.

Overt Observer

An overt observer assumes a passive role during interactions, but may approach participants post hoc for interviews. The effect of the presence of an observer is largely unclear, there might be problems with making users feel uncomfortable when they are approached post hoc and realize that they were part of a study.

Covert Observer

A covert observer is not visible to the participants and assumes no active role during the complete process. Thus, data can solely be collected through video and audio recordings as well as interaction logs from the system.

Williamson and Williamson [32] set up a field study to quantify the effect of different roles to gain more insight in their influence on the collected data. They uncover a difference in interaction distance depending on the observer role. Moreover, they found a significant difference in converting passers-by into active users of the system, with covert observer resulting in significantly higher conversion rate.

These results come from a public evaluation with uninformed users. This is not necessarily the same situation when we deploy a socially assistive agent system in a health care setting. Here, the system might be used by a limited number of informed users. But it shows that the role of the experimenter has a quantifiable effect and should be carefully taken into account in planning a field study.

4.2.3 Controlled Versus Uncontrolled Field Studies

Another distinction can be made by the way participants are recruited field studies. Claes and colleagues [5] call this controlled versus uncontrolled studies, where in the first case, participants are recruited and invited to participate in the study that takes place at the deployment site, and in the latter case, participants are passers-by that interact voluntarily with the system.

Their results show advantages and disadvantages for both approaches. They looked at the following variables: passive engagement (noticing/looking at system), active engagement (interacting with system), insight depth (number of information gained from system), and social interaction (interaction with others/honeypot effect).

No statistical difference was found between the two conditions, but more people participated in the uncontrolled study. The controlled study on the other hand delivered qualitatively better results, e.g., participant tried out more interactions.

For agent systems in institutional and home care settings, we can assume a controlled field deployment, where users and support staff are carefully selected. Based on the insights presented here, we can assume qualitatively good results but the study setting itself results naturally in a low sample size.

4.2.4 Role Fluidity During Field Studies

Boyd and colleagues [2] emphasize that field deployments are coordinated endevours that require the active participation of researchers, participants, stakeholders, as well as supporters, e.g., in the form of care personnel or relatives when in comes to assistive technologies. One important feature is the so-called articulation work to support the use of new technologies and to guide users in engaging with the system. While a well-known fact in HCI research, it is rarely acknowledged in work around socially assistive agent technology, which is supposed to work autonomously and self-explanatory (intuitive).

In their paper, Boyd and colleagues show how this articulation work in a field study of collaborative tablet games is facilitated by role changes or rather role enhancements of the different participants, where the researchers become community collaborators, the students become research participants, and the teachers become research facilitators.

In our work on co-creation of personal reminder robots with people with acquired brain damage [24], we have exploited this concept of role fluidity for the entire research process, where target users as well as their primary support staff became members of the development team, while at the same time keeping their role as users. In order to test agent technology in care setting, it will be indispensable to rely on the support and ownership of the care personnel and/or relatives to ensure a successful field deployment. Especially considering the fact that something will go wrong and will require adjustments from the research team. Thus, frustration on the side of the user has to be actively countered.

4.3 Methods for Field Studies

The HCI research presented in the previous section investigated public evaluations, i.e., field studies with a large number of user, usually passers-by, in public spaces. With multimodal and multicultural agent systems, we aim for a different context, which is usually a controlled space, e.g., in an institutional or home setting. Thus, some of the external factors like weather or vandalism might be of less importance.

One caveat right at the beginning: Field studies are expensive in terms of time and equipment and travel. Thus, when we deploy a system in the field, we would like to capture as much data as possible. In order to capture as much data as possible, a mix of methods is necessary for data collection:

- Interaction logs: Every system that is used in a field study should log usage of the system. This will provide in depth information, e.g., on which services the user selected or how long and how often s/he engaged with the system (e.g., [7]).
- Sensor data: Multimodal agent system are equipped with a range of sensor capabilities to capture user input and interpret the environment. This can be vision based techniques with rgb or depth camera. It can also be physiological or even BCI data. Saving the data for alter analysis provides a detailed picture of the interactions with the system and might allow for automatic annotation of interaction patterns (e.g., [30])
- Observations: A rich source of information is video recordings and field notes of the user interacting with the system. In contrast to sensor data mentioned above, observations are analyzed manually, making use of grounded theory as a method to discover patterns in the observations (e.g., [18]).
- Interviews: Another qualitative method is interviews after the user has interacted with the system. With the help of interviews, we can reach an understanding of the users opinions towards the system, reasons for specific behaviors and actions as well as the users experiences with and feeling towards the system (e.g., [3]).
- Questionnaires: A standard tool, also in lab studies, is questionnaires, which are used to capture subjective impressions about the interaction with the system (e.g., [16]). In field studies, it might be more difficult to administer the questionnaire, if they are not integrated in the interaction with the user.

Examples of Mixed Method Field Studies

Mueller and colleagues [17] investigate interactions with a shop window. They combine interaction logs for person tracking with videos from depth camera, which allows automatic detection of interactions. Additional cameras are used for video observations by a covert observer. Both camera feeds are manually coded to capture interaction patterns. The data was complemented by semi-structures interviews with stakeholders, i.e., the shop owners and selected users.

Morrison and colleagues [15] present the Humming Wall, an interactive art installation in an urban environment. Their main data collection tool are thermal cameras that allow capturing user interactions without privacy violations and the need for informed consent (see also below Sect. 4.4). The video streams have been coded manually for interaction time, interaction activity, and group composition, i.e., individual users, pairs of users, more than two users, and children.

Palacin-Silva and colleagues [20] compare game-based and non-game-based interaction with a mobile app for lake monitoring in a controlled field study with convenience sampling from university students. The system's interaction logs capture number of submissions, number of active users for the whole duration of the trial, number of incomplete trials as well as time to learn the app during the first usage. This data is accompanied by a battery of pre- and post questionnaires such as SUS (system usability scale), IBM computer usability satisfaction, user acceptance of information technology, and perceived playfulness.

Yang and colleagues [33] investigate gesture-based smartphone authentication. In order to create realistic test scenario, they use the experience sampling method. Participants need to perform tasks at different times during their days. The app will notify them when it is time to perform a task. The interaction logs give information about the number of created password gestures as well as the number of completed log-ins. The study ends with a NASA TLX questonaire for measuring subjective task load for each participant.

Craven and colleagues [6] highlight the benefits of large scale field deployments that allow for collecting real user data through data logging in an educational game for a large public display. The report on a five month long test with data logging resulting in 781 plays, including questionnaires for subjective ratings after each play through. They use a self made questionnaire with four questions in relation to the user experience.

Vaizman and colleagues [29] report on a mobile data collection app that allows to continuously collect sensor data and offers the possibility of labeling this data by the user. This allows for crowdsourcing data collection and annotation for training data bases for machine learning approaches. Users have to be aware that they are constantly monitored and that their data is send to a central server.

Pang and colleagues [21] present four week field study of city explorer, a location-based mobile app. Twelve participants were recruited to use the app during their daily commutes. Interaction logs included number and duration of game sessions as well as dates and times of game plays. The logs were accompanied by pre- and post semi-structured Interviews that were analyzed using open coding. Additionally, participants received weekly email surveys with ten questions.

Bianchi-Berthouze and colleagues [1] have a more qualitative approach to data collection and focus on the analysis of online material, interviews, and observations. This allow getting insights into the experiential structures of the phenomenon under observation, here long distance running, and can lead to design and development guidelines for specifically targeted applications.

The examples presented here are thought as inspiration and highlight the different possibilities for selecting methods for data collection in a field study. Apart from the type of information that should be obtained from the field study, familiarity with methods can also be a selection criterion. Trying out a new method for the first time in the field study is not a good idea due to the time and cost that will be put into the study. Thus, for new methods, there should always be a training phase before the field deployment.

4.4 Ethical Concerns for Field Studies

Running field studies, especially in health care scenarios, there are a number of ethical considerations in relation to who we test with, which and how data is collected, or what happens when the system is removed by the end of the study. Some of these have been formalized, e.g., through ethical boards, others result from the inherent characteristics of the technology that we enter into the field.

Formal ethical requirements depend on the country in which the study is performed. Especially work with vulnerable people such as users in health care contexts often requires an authorization from an ethical committee. The specific rules and the amount of documentation depends on the local legislation.[1]

Typically, informed consent has be collected from all participants and/or their legal guardians. Participants need to be informed about the goals of the project, which data is collected for which purpose. If there is an intention of using selected data in publications, the participant needs to agree to this. Multimodal agent systems pose the challenge that they use sophisticated sensor technology to enable intuitive or natural interactions with the user, thus they will record visual, auditory or even physiological data from the user. If this kind of data is collected, the user needs to be informed about the type of data, its use in and outside the system (e.g., for further research purposes or publications), its storage, and its expiry date. Especially the last point presents an ethical paradox. At least according to the European legislation,[2] data can only be collected for a specified research project and a date close to the end of the research project has to be given when the data will be deleted. But the data is used in scientific research, thus deleting the data makes it impossible to replicate or falsify any conclusions drawn from analyzing the data. Also, sharing of data with other researchers becomes difficult, a common practice, for example, in the machine learning field.

An additional challenge in the context of institutional or home care is the possibility that there are several people in the environment (care personnel, cleaning staff, relatives, etc.) [25]. In principle, we need informed consent from all of them, which might be difficult, especially when the system is used in a long-term study.

Other ethical considerations are less formalized but might present issues during the field study regarding the adoption and use of the agent system. To ensure this, the agent's behavior and actions should be transparent and explainable. We have seen above the need for articulation work from the secondary users (like teacher or care personnel) or the researchers to support the continuous use of the system in field studies. In case of an agent system, it might sometimes be possible that the system itself can take over some of the articulation work by providing motivations for actions.

Multimodal agent or robot systems provide the unique possibility to create relations and emotional bonds with the user (e.g., [4, 13]). In principle, this is a positive

---

[1] On an EU level, there is an increased awareness of real-life experimentation; see, e.g., https://cordis.europa.eu/project/id/872441.

[2] https://gdpr.eu/.

asset in the envisioned scenarios. But if the user has problem remembering and maybe forgets how the system is supposed to work, it might also induce negative feelings of shame or frustration that might result in under- or misuse of the system. Also, the agent system might be exhibiting normative behavior that is in line with the policies of the care institution but might add odds with the user's interests. An example from a recent ethnographic study we conducted in a 24/7 care facility for people with acquired brain damage, for example, showed a tension in shopping routines between the institution's care policy regarding healthy eating habits and the residents' wish for sweets and alcoholic beverages. A potential shopping assistant will thus be working in this tension between norms and regulations on the one hand and the user wishes and independence on the other hand, which might result in uncooperative behavior of the system at least from the user's perspective.

If the system proves to be useful and is adopted by the participants and their surroundings, then we face an additional challenge at the end of the field study. Participants will become attached to the system and it will be problematic to withdraw the system at the end of the study. Participants need to be aware from the beginning that the system is only temporarily available. Even if the system could be left with the participant, we as researchers could never guarantee support and maintenance that goes beyond the time frame of the field study. There is no good solution to this problem and we have to be aware of the emotionally challenging situation this can lead to. In our work, we found it helpful to mark the end of a field study with a little festive event, where we give an account of the project and recognize the engagement of the participants. Thus, there is a clear end of the field study for both participants, environment, and the researchers.

4.5 Conclusion or Is It Worth the Effort?

This chapter looked at considerations for evaluating multimodal agent systems outside of the lab. In principle, field research should actually start much earlier in order to understand the context for which the agent is developed and to understand the practices and routines into which the agent will be integrated (e.g., [24]). Starting early with field research has the additional advantage of getting to know relevant stakeholders and potential participants for later field studies with the system.

Field tests allow us to test agents under realistic conditions in the actual context of use with real users. Lab research can bias results in several ways. We cannot recreate the proper environmental factors that play major roles in the real context and we might not be able to test all tasks because often they will be too constrained and unrealistic under lab conditions. To give an example, reminders are often used as prototypical examples for socially assistive robots (e.g., [9, 12, 14, 22]), but it is nearly impossible to test reminders in the lab because, we cannot create situations where a reminder is necessary or might make sense.

When we develop multimodal multicultural agents for institutional or home care contexts, we also assume that users will interact repetitively and over a longer time with the agent. Ideally, the user builds a relation to the agent touching upon factors like trust and emotions. Again, such long- term studies only make sense if they are run as field studies, because controlled experimental study can give us only a fraction of the information that we need to for these challenges.

References

1. Bi, T., Bianchi-Berthouze, N., Singh, A., Costanza, E.: Understanding the shared experience of runners and spectators in long-distance running events. In: Proceedings of the 2019 CHI Conference on Human Factors in Computing Systems, CHI '19, pp. 461:1–461:13. ACM, New York, NY, USA (2019). https://doi.org/10.1145/3290605.3300691
2. Boyd, L.E., Rector, K., Profita, H., Stangl, A.J., Zolyomi, A., Kane, S.K., Hayes, G.R.: Understanding the role fluidity of stakeholders during assistive technology research "in the wild". In: Proceedings of the 2017 CHI Conference on Human Factors in Computing Systems, CHI '17, pp. 6147–6158. ACM, New York, NY, USA (2017). https://doi.org/10.1145/3025453.3025493
3. Brinkmann, S., Kvale, S.: InterViews: Learning the Craft of Qualitative Research Interviewing. SAGE Publications, California (2014)
4. Charles, F., Pecune, F., Aranyi, G., Pelachaud, C., Cavazza, M.: Eca control using a single affective user dimension. In: Proceedings of the 2015 ACM on International Conference on Multimodal Interaction, ICMI '15, pp. 183–190. Association for Computing Machinery, New York, NY, USA (2015). https://doi.org/10.1145/2818346.2820730
5. Claes, S., Wouters, N., Slegers, K., Vande Moere, A.: Controlling in-the-wild evaluation studies of public displays. In: Proceedings of the 33rd Annual ACM Conference on Human Factors in Computing Systems, CHI '15, pp. 81–84. ACM, New York, NY, USA (2015). https://doi.org/10.1145/2702123.2702353
6. Craven, M.P., Simons, L., Gillott, A., North, S., Schnädelbach, H., Young, Z.: Evaluating a public display installation with game and video to raise awareness of attention deficit hyperactivity disorder. In: Kurosu, M. (ed.) Human-Computer Interaction: Interaction Technologies, pp. 584–595. Springer International Publishing, Cham (2015)
7. Dumais, S., Jeffries, R., Russell, D.M., Tang, D., Teevan, J.: Understanding User Behavior Through Log Data and Analysis, pp. 349–372. Springer, New York (2014). https://doi.org/10.1007/978-1-4939-0378-8_14
8. Fischer, P.T., Hornecker, E.: Urban HCI: spatial aspects in the design of shared encounters for media facades. In: Proceedings of the SIGCHI Conference on Human Factors in Computing Systems, CHI '12, pp. 307–316. Association for Computing Machinery, New York, NY, USA (2012). https://doi.org/10.1145/2207676.2207719
9. Khan, A., Anwar, Y.: Robots in healthcare: a survey. In: Arai, K., Kapoor, S. (eds.) Advances in Computer Vision, pp. 280–292. Springer International Publishing, Cham (2020)
10. Mäkelä, V., Sharma, S., Hakulinen, J., Heimonen, T., Turunen, M.: Challenges in public display deployments: A taxonomy of external factors. In: Proceedings of the 2017 CHI Conference on Human Factors in Computing Systems, CHI '17, pp. 3426–3475. ACM, New York, NY, USA (2017). https://doi.org/10.1145/3025453.3025798
11. Mariager, C.S., Fischer, D.K.B., Kristiansen, J., Rehm, M.: Co-designing and field-testing adaptable robots for triggering positive social interactions for adolescents with cerebral palsy. In: 2019 28th IEEE International Conference on Robot and Human Interactive Communication (RO-MAN), pp. 1–6. IEEE Press (2019)
12. Martinez-Martin, E., del Pobil, A.P.: Personal Robot Assistants for Elderly Care: An Overview, pp. 77–91. Springer International Publishing, Cham (2018). https://doi.org/10.1007/978-3-319-62530-0_5

13. Miyamoto, T., Katagami, D., Shigemitsu, Y.: Improving relationships based on positive politeness between humans and life-like agents. In: Proceedings of the 5th International Conference on Human Agent Interaction, HAI '17, pp. 451–455. Association for Computing Machinery, New York, NY, USA (2017). https://doi.org/10.1145/3125739.3132585
14. Mois, G., Beer, J.M.: The Role of Healthcare Robotics in Providing Support to Older Adults: a Socio-ecological Perspective. Current Geriatrics Reports (2020)
15. Morrison, A., Manresa-Yee, C., Jensen, W., Eshraghi, N.: The humming wall: vibrotactile and vibroacoustic interactions in an urban environment. In: Proceedings of the 2016 ACM Conference on Designing Interactive Systems, DIS '16, pp. 818–822. ACM, New York, NY, USA (2016). https://doi.org/10.1145/2901790.2901878
16. Müller, H., Sedley, A., Ferrall-Nunge, E.: Survey Research in HCI, pp. 229–266. Springer, New York (2014). https://doi.org/10.1007/978-1-4939-0378-8_10
17. Müller, J., Walter, R., Bailly, G., Nischt, M., Alt, F.: Looking glass: A field study on noticing interactivity of a shop window. In: Proceedings of the SIGCHI Conference on Human Factors in Computing Systems, CHI '12, pp. 297–306. ACM, New York, NY, USA (2012). https://doi.org/10.1145/2207676.2207718
18. Muller, M.: Curiosity, Creativity, and Surprise as Analytic Tools: Grounded Theory Method, pp. 25–48. Springer New York, New York, NY (2014). https://doi.org/10.1007/978-1-4939-0378-8_2
19. Norouzi, N., Kim, K., Hochreiter, J., Lee, M., Daher, S., Bruder, G., Welch, G.: A systematic survey of 15 years of user studies published in the intelligent virtual agents conference. In: Proceedings of the 18th International Conference on Intelligent Virtual Agents, IVA '18, pp. 17–22. Association for Computing Machinery, New York, NY, USA (2018). https://doi.org/10.1145/3267851.3267901
20. Palacin-Silva, M.V., Knutas, A., Ferrario, M.A., Porras, J., Ikonen, J., Chea, C.: The role of gamification in participatory environmental sensing: A study in the wild. In: Proceedings of the 2018 CHI Conference on Human Factors in Computing Systems, CHI '18, pp. 221:1–221:13. ACM, New York, NY, USA (2018). https://doi.org/10.1145/3173574.3173795
21. Pang, C., Pan, R., Neustaedter, C., Hennessy, K.: City explorer: The design and evaluation of a location-based community information system. In: Proceedings of the 2019 CHI Conference on Human Factors in Computing Systems, CHI '19, pp. 341:1–341:15. ACM, New York, NY, USA (2019). https://doi.org/10.1145/3290605.3300571
22. Petrie, H., Darzentas, J.: Older people and robotic technologies in the home: perspectives from recent research literature. In: Proceedings of the 10th International Conference on PErvasive Technologies Related to Assistive Environments, pp. 29–36. Association for Computing Machinery, New York, NY, USA (2017). https://doi.org/10.1145/3056540.3056553
23. Rehm, M., Jensen, M.L.: Accessing cultural artifacts through digital companions: the effects on children's engagement. In: 2015 International Conference on Culture and Computing (Culture Computing), pp. 72–79. IEEE Press (2015)
24. Rehm, M., Rodil, K., Krummheuer, A.L.: Developing a new brand of culturally-aware personal robots based on local cultural practices in the danish health care system. In: 2018 IEEE/RSJ International Conference on Intelligent Robots and Systems (IROS), pp. 2002–2007 (2018). https://doi.org/10.1109/IROS.2018.8594478
25. Reig, S., Luria, M., Wang, J.Z., Oltman, D., Carter, E.J., Steinfeld, A., Forlizzi, J., Zimmerman, J.: Not some random agent: Multi-person interaction with a personalizing service robot. In: Proceedings of the 2020 ACM/IEEE International Conference on Human-Robot Interaction, HRI '20, pp. 289–297. Association for Computing Machinery, New York, NY, USA (2020). https://doi.org/10.1145/3319502.3374795
26. Rodil, K., Rehm, M., Krummheuer, A.L.: Co-designing social robots with cognitively impaired citizens. In: Proceedings of the 10th Nordic Conference on Human-Computer Interaction, NordiCHI '18, pp. 686–690. ACM, New York, NY, USA (2018). https://doi.org/10.1145/3240167.3240253
27. Siek, K.A., Hayes, G.R., Newman, M.W., Tang, J.C.: Field Deployments: Knowing from Using in Context, pp. 119–142. Springer, New York (2014). https://doi.org/10.1007/978-1-4939-0378-8_6

28. Ssozi-Mugarura, F., Reitmaier, T., Venter, A., Blake, E.: Enough with "in-the-wild". In: Proceedings of the First African Conference on Human Computer Interaction, AfriCHI '16, pp. 182–186. Association for Computing Machinery, New York, NY, USA (2016). https://doi.org/10.1145/2998581.2998601
29. Vaizman, Y., Ellis, K., Lanckriet, G., Weibel, N.: Extrasensory app: data collection in-the-wild with rich user interface to self-report behavior. In: Proceedings of the 2018 CHI Conference on Human Factors in Computing Systems, CHI '18, pp. 554:1–554:12. ACM, New York, NY, USA (2018). https://doi.org/10.1145/3173574.3174128
30. Voida, S., Patterson, D.J., Patel, S.N.: Sensor Data Streams, pp. 291–321. Springer, New York (2014). https://doi.org/10.1007/978-1-4939-0378-8_12
31. Voit, A., Mayer, S., Schwind, V., Henze, N.: Online, vr, ar, lab, and in-situ: Comparison of research methods to evaluate smart artifacts. In: Proceedings of the 2019 CHI Conference on Human Factors in Computing Systems, CHI '19, pp. 507:1–507:12. ACM, New York, NY, USA (2019). https://doi.org/10.1145/3290605.3300737
32. Williamson, J.R., Williamson, J.: Understanding public evaluation: quantifying experimenter intervention. In: Proceedings of the 2017 CHI Conference on Human Factors in Computing Systems, CHI '17, pp. 3414–3425. ACM, New York, NY, USA (2017). https://doi.org/10.1145/3025453.3025598
33. Yang, Y., Clark, G.D., Lindqvist, J., Oulasvirta, A.: Free-form gesture authentication in the wild. In: Proceedings of the 2016 CHI Conference on Human Factors in Computing Systems, CHI '16, pp. 3722–3735. ACM, New York, NY, USA (2016). https://doi.org/10.1145/2858036.2858270

Chapter 5
Socio-Cognitive Language Processing for Special User Groups

Björn W. Schuller and Michael F. McTear

5.1 Introduction

Sociocognitive Language Processing (SCLP)[1] can be considered as a term that covers 'soft factors' in communication. Likewise, it can be seen as a specific kind of and arguably also as an extension to the more traditional and broadly defined field of Natural Language Processing (NLP)—the idea of coping with everyday language, including slang and multi-lingual phrases and cultural aspects, and in particular, with irony/sarcasm/humour, as well as paralinguistic information such as the physical and mental state and traits of the dialogue partner (e.g., affect, age groups, personality dimensions), and social aspects. Additionally, multimodal aspects such as facial expression, gestures or bodily behaviour should ideally be included in the analysis wherever possible. At the same time, SCLP can render future dialogue systems more 'chatty' by not only appearing natural but also by being truly emotionally and socially competent, ideally leading to a more symmetrical dialogue. To do this, the computer should itself have a 'need for humour', an 'increase of familiarity', etc., i.e., enabling computers to experience or at least better understand emotions and personality so that that they have 'a feel' for these concepts. Beyond these ideas, the broader idea of SCLP includes verbal behaviour analysis, a closer coupling between language understanding and generation incorporating social and affective information, and new language resources to meet these ends. In this way, SCLP links NLP expertise more closely with that of psychology, the social sciences, and related disciplines.

[1] As the field of *Spoken* Language Processing is usually abbreviated as SLP, we suggest S*C*LP as a short notation for Sociocognitive Language Processing.

B. W. Schuller (✉)
GLAM – Group on Language, Audio, & Music, Imperial College London, London, UK
e-mail: bjoern.schuller@imperial.ac.uk

M. F. McTear
Computer Science Research Institute, Ulster University, Northern Ireland, UK
e-mail: mf.mctear@ulster.ac.uk

J. Miehle et al. (eds.), *Multimodal Agents for Ageing and Multicultural Societies*,
https://doi.org/10.1007/978-981-16-3476-5_5

In this short paper, we will exemplify the principle by focusing on the analysis side of NLP, known as Natural Language Understanding (NLU). Further, we will limit the example to *spoken* language understanding (SLU). The principles do, however, similarly apply to Natural Language Generation (NLG) in any form.

5.2 Spoken Language Understanding

In SLU, the text output by a speech recognition system is analyzed in order to determine its meaning. This meaning representation can then be used in a spoken dialogue application or in other tasks such as speech mining, speech information retrieval, or speech translation.

Three stages can be distinguished in the development of computational approaches to language understanding. Up until the late 1980s, developers handcrafted grammars that consisted of rules covering all predicted inputs along with a parser that applied the grammar rules to the input to determine its constituent structure. In this approach, the focus was on the analysis of written texts i.e. natural language understanding (NLU). In the late 1980s, a paradigm shift occurred in which probabilistic and data-driven models that had already been deployed successfully in speech recognition were now applied to language understanding. In this approach, attention turned to spoken language understanding (SLU), as the statistical methods were more able to deal with the ill-formed input typical of spontaneous speech. By around 2006, a new approach was emerging with the application of deep learning neural models to language understanding and the use of end-to-end architectures that eliminated the need for the traditional components of pipelined architectures.

5.2.1 Rule-Based Approaches

In theoretical computational linguistics, language understanding involved two stages of analysis:

1. *Syntactic analysis*—to determine the constituent structure of the input.
2. *Semantic analysis*—to determine the meanings of the constituents.

This two-step approach is based on the *principle of compositionality*, which states that the meaning of a complex expression is determined by the meanings of its constituent parts and the rules used to combine them. In this approach, fine-grained distinctions can be captured through a deep analysis of constituent structure that has a direct bearing on the semantic analysis. For example, there is only one word that is different in the following sentences, but changing the words results in different meaning representations:

- S1: *List all employees of the companies who are based in the city centre*
- S2: *List all employees of the companies that are based in the city centre*

The interpretation of S1 asks for a listing of employees who are based in the city centre while the interpretation of S2 asks for a listing of employees who are not necessarily based in the city centre, but who work for companies based there. This difference can only be picked up by an analysis that reflects the difference between the use of *who* and *that* in these sentences.

Following the syntactic analysis, the syntactic constituents are transformed into a meaning representation that typically takes the form of a logic-based formalism. Using logic in this way offers a deeper level of understanding, as it enables the application of standard mechanisms for inference. For a detailed account of the formalisms used in rule-based NLU, see [14].

In an alternative approach, semantic analysis is performed directly on the input using a *semantic grammar*. Although this approach is not theoretically motivated, it has been applied successfully in dialogue systems to analyze the user's inputs. Semantic grammars are more robust to the sorts of ungrammatical input and recognition errors that occur in spontaneous speech as they focus on the keywords in the input and do not have to analyze every word. However, they are usually domain-specific, so that separate grammars are required for each new domain.

5.2.2 Statistical Approaches

With the emergence of spoken dialogue systems in the late 1980s and early 1990s, it became apparent that the rule-based approaches used extensively in NLU were not sufficient for spoken language understanding (SLU). Whereas the input to NLU was well-formed written text, in spoken dialogue systems, SLU was required to analyze spoken text that did not necessarily follow the same grammar rules, as it often contained self-corrections, hesitations, repetitions, and other types of dysfluency. Moreover, the output of a speech recognition system took the form of a stream of tokens with no structure information such as punctuation to help determine sentence boundaries and structure within the sentence. In the rule-based approach, the input has to match the rules exactly and any small variations require additional rules in order to be accepted by the grammar. In this respect, the statistical approach is more robust as it can handle input that is potentially ill-formed as well as synonymous strings that should result in the same meaning representation, but would require additional rules in the rule-based approach.

In statistical SLU, the focus was mainly on analyzing the language produced in domain-specific task-based dialogue systems—for example, flight reservations, hotel bookings, etc. With the increasing availability of large corpora of spoken dialogues, it became possible to use machine learning techniques to automatically learn mappings between spoken inputs and the required meaning representations from labelled training data. Given the nature of the tasks, it was not necessary to derive a logic-based meaning representation. Instead, the output of SLU in a spoken dialogue system is typically a frame-based representation consisting of sets of attribute-value pairs that capture the information in an utterance that is relevant to the application [17]. Generally, three elements are extracted: the *domain* to which the utterance relates (for example, flight reservations), the user's *intent* (for example, to book a flight), and the *entities* that are required to make the reservation (for example, **destination**, **date of travel**, etc.). Thus, the representation of an utterance such as *book a flight to London on Friday* would be something like: *domain = flight_reservations*; *intent = book_flight*; *entities: destination = London, day = Friday*. These elements are extracted from the input using machine learning methods such as classifiers (e.g., support vector machines) for the domain and intent, and Conditional Random Fields for the entities [38].

5.2.3 Deep Learning Approaches

Since around 2006, deep learning techniques have been applied in NLU and have been shown to outperform other machine learning-based approaches. For example, [17] used Recurrent Neural Networks for the identification of intents and the extraction of entities, while [13] used a bi-directional RNN to jointly classify intents and extract entities.

Deep learning for SLU makes use of the sequence-to-sequence (Seq2Seq) model, in which, the input and the output are represented as a sequence [33, 36]. The model consists of an encoder and a decoder. The encoder processes the input creating a vector known as a context (or thought) vector that represents the final hidden state of the encoder. The decoder takes this vector and uses it to create the output. Seq2Seq has been used in machine translation to perform a transduction from an input in a source language such as English to an output in a target language such as German, and in dialogue systems to perform a transduction from an input utterance to a response. Encoding involves the use of neural networks, usually recurrent neural networks (RNNS), long short-term memory networks (LSTMs), gated recurrent units (GRUs), and more recently transformer networks. Decoding takes one element at a time to produce an output sequence using the context vector where the word that is generated at each time step is conditioned on the word generated by the network at the previous time step, as well as the hidden state, providing the context from the previous time step.

There are many variations on the encoder-decoder architecture. For example, the *attention mechanism* was introduced to address the problem that performance decreases as the input sequence becomes longer [3]. The more recently introduced Transformer architecture uses attention mechanisms to draw global dependencies between the input and the output [40]. Whereas a traditional encoder treats every item in a sequence as equally relevant, the transformer selects which parts to include in the encoding as a basis for the current prediction. For more details on deep learning in SLU, see [37].

5.2.4 Implications for Socio-Cognitive Language Understanding

Currently, SLU focuses primarily on analyzing the textual form of a message and on outputting a representation of its propositional content. However, additional information is conveyed in the prosodic features of a spoken utterance—its phrasing, pitch, loudness, tempo, and rhythm—that can indicate differences in the function of an utterance as well as expressing emotional aspects such as anger or surprise [34]. Other information to support the interpretation of an utterance may come from sensors that provide data about the environmental context, biosensors that report on the user's physical and emotional state, and machine vision systems that can detect non-verbal accompaniments of speech, such as gestures and facial expressions.

Recently researchers in neural dialogue have started to explore how to integrate information about emotional aspects into their models. Ghosh et al. [11] present an extension to an LSTM language model for generating conversational text that was trained on conversational speech corpora. The model predicts the next word in the output conditioned not only on the previous words, but also on an affective category that infers the emotional content of the words. In this way, the model is able to generate expressive text at various degrees of emotional strength. Zhou et al. [42] developed a conversational model called *Emotional Chatting Machine (ECM)* that produces emotion-based responses to any user input. See also [1] for discussion of a study, in which affective content was incorporated into LSTM encoder-decoder neural dialogue models enabling them to produce emotionally rich and more interesting and natural responses.

Adding additional multimodal input streams and integrating them, if required, into a single meaning representation presents challenges that have yet to be addressed. Similarly, generating output that makes use of this richer information is beyond the current state of the art. We discuss the challenges for Sociocognitive Language Processing of adopting these additional features in the next section.

5.3 The Sociocognitive View

Above (cf. Sect. 5.2), we introduced SLU and showed that, in principle, works in this field *do* consider prosody, and often also multimodal information such as considering the facial expression of speakers. One may thus ask what makes the term SCLP different or justified. In [12], the authors find in two experiments that it seems plausible to consider language understanding as "a special case of social cognition". This is based on a model to predict an "interaction between the speaker's knowledge state and the listener's interpretation" [12]. Similarly, the authors in [9] attest the high relevance of "social, cognitive, situational, and contextual aspects" when dealing with language. Further, the fields of Affective Computing (AC) [25], Social Signal Processing (SSP) [24, 41] or Behavioural Signal Processing (BSP) [20], suggest consideration of affective or emotional and social cues for computing systems used, e.g., in human–computer interaction [21], or human dialogue analysis. In particular, in speech analysis [2, 7, 28, 30] and synthesis [18, 19], such information was considered rather early. Further, sub-fields of NLP deal with such information, most noteworthy the discipline of Sentiment Analysis (SA) [6, 15, 22, 32, 39]. Including also acoustic speech feature information, Computational Paralinguistics (CP) [31] provides a broader view on speaker states and traits beyond sentiment, emotion, or social signals including also biological traits such as age, gender, height, or race, personality traits, or health-related state and trait information (cf. also [10, 23]) alongside physiological states such as eating or exercising, and cognitive load.

Likewise, AC and SSP/BSP provide a general beyond-language paradigm for computer analysis and synthesis of affective and behavioural cues—each of which focusing on one of emotional or social intelligence; SA (and related sub-disciplines of NLP such as Opinion Mining) focus on the analysis of a single aspect—here sentiment—and neglect the synthesis side, and CP deals mainly with speech and language analysis and synthesis without a linking model or component such as a dialogue model. This makes NLP the definition that comes arguably closest to SCLP; however, while it aims to deal with 'natural' language, its emphasis on the soft factors of communication is rather weak. By SCLP, we advocate a strong link between language and the authoring or speaking person's state and trait as related to the spoken content and conveyed 'message'. Likewise, a decision such as in the example in Sect. 5.2 regarding the interpretation of S1 potentially in the sense of S2 could be supported by estimation of the social and cultural background of the speaker such as 'native speaker' (or not) or the personality such as 'conscientious' (then, likely taking it for the actual sense of S1).

5.4 Conclusion

In this short contribution, we introduced the term of Sociocognitive Language Processing (SCLP). We further motivated its introduction reviewing key relevant literature in a constructive and synthetic manner with the aim to highlight borders between related existing disciplines and terms such as NLP and SA, and at the same time confine SCLP as compared to broader fields such as AC and SSP.

We exemplified the considerations by language analysis looking at NLU and leading to a sociocognitive view. In the same vein, in NLG, the sociocognitive view lends more weight to the 'soft factors' in communication, such as synthesizing irony, or non-verbal fillers, and behaviours that fit these and accompany the linguistic content. We believe that, by integrating SCLP principles, one can render future dialogue systems more 'chatty' making them not only feel 'natural', but truly emotionally and socially competent (cf., e.g., [8, 29]). Ideally, this will lead to a more 'symmetrical' dialogue, where both ends—humans and computer systems or intelligent machines—will integrate and comprehend soft factors in the communication.

Arguably, for that, communicative technical systems should themselves have a 'need for humour' [5, 26, 35], an 'increase of familiarity' during repeated or prolonged interactions, etc. In other words, it appears required for genuine SCLP to enable computers to experience and have or at least better understand emotions and personality, such that they have 'a feel' for these concepts (cf., e.g., [4]). For example, the degree of conscientiousness of a system might be the decisive factor between taking S1 in our example in the sense of S1 or rather S2 in addition to its interpretation of 'what the user is like' (based also on increased familiarity) in relation to the best interpretation. This will, however, need to be further expanded upon in follow-up considerations and studies including 'Sociocognitive Dialogue Processing'. Beyond these ideas, the broader idea of SCLP includes verbal behaviour analysis, a closer coupling between language understanding and generation incorporating social and affective information, and new language resources to meet these ends. By that, SCLP unites expertise from psychology and social sciences with NLP on the way to enable genuine conversational dialogue systems [16] or emotionally and socially aware computer-mediated communication [27].

Acknowledgements The first author acknowledges funding from the ERC under grant agreement no. 338164 (iHEARu), and the European Union's Horizon 2020 Framework Programme under grant agreements nos. 645378 (ARIA-VALUSPA), 644632 (MixedEmotions), and 645094 (SEWA).

References

1. Asghar, N., Poupart, P., Hoey, J., Jiang, X., Mou, L.: Affective neural response generation. In: European Conference on Information Retrieval, pp. 154–166. Springer (2018)
2. Bachorowski, J.A., Owren, M.J.: Vocal expression of emotion: acoustic properties of speech are associated with emotional intensity and context. Psychol. Sci. **6**(4), 219–224 (1995)
3. Bahdanau, D., Cho, K., Bengio, Y.: Neural machine translation by jointly learning to align and translate (2014). arXiv:1409.0473
4. Ball, J.E., Breese, J.S.: Modeling and projecting emotion and personality from a computer user interface (2001). US Patent 6,212,502
5. Binsted, K., Bergen, B., O'Mara, D., Coulson, S., Nijholt, A., Stock, O., Strapparava, C., Ritchie, G., Manurung, R., Pain, H., Waller, A., O'Mara, D.: Computational humor. IEEE Intell. Syst. **21**(2), 59–69 (2006)
6. Cambria, E., Schuller, B., Xia, Y., Havasi, C.: New avenues in opinion mining and sentiment analysis. IEEE Intell. Syst. Mag. **28**(2), 15–21 (2013)
7. Dellaert, F., Polzin, T., Waibel, A.: Recognizing emotion in speech. In: Proceedings of the Fourth International Conference on Spoken Language Processing, vol. 3, pp. 1970–1973. IEEE (1996)
8. DeVault, D., Artstein, R., Benn, G., Dey, T., Fast, E., Gainer, A., Georgila, K., Gratch, J., Hartholt, A., Lhommet, M., Lucas, G., Marsella, S., Morbini, F., Nazarian, A., Scherer, S., Stratou, G., Suri, A., Traum, D., Wood, R., Xu, Y., Rizzo, A., Morency, L.P.: Simsensei kiosk: a virtual human interviewer for healthcare decision support. In: Proceedings of the 2014 International Conference on Autonomous Agents and Multi-agent Systems, pp. 1061–1068. International Foundation for Autonomous Agents and Multiagent Systems (2014)
9. Dietrich, R., Graumann, C.F.: Language Processing in Social Context. Elsevier, Amsterdam (2014)

10. Furnham, A.: Language and personality. In: Giles, H., Robinson, P. (eds.) Handbook of Language and Social Psychology. Wiley, Hoboken (1990)
11. Ghosh, S., Chollet, M., Laksana, E., Morency, L.P., Scherer, S.: Affect-lm: a neural language model for customizable affective text generation (2017). arXiv:1704.06851
12. Goodman, N.D., Stuhlmüller, A.: Knowledge and implicature: modeling language understanding as social cognition. Top. Cogn. Sci. **5**(1), 173–184 (2013)
13. Hakkani-Tür, D., Tür, G., Celikyilmaz, A., Chen, Y.N., Gao, J., Deng, L., Wang, Y.Y.: Multi-domain joint semantic frame parsing using bi-directional rnn-lstm. In: Interspeech, pp. 715–719 (2016)
14. Jurafsky, D., Martin, J.H.: Speech and Language Processing (3rd ed. draft) (2020). https://web.stanford.edu/~jurafsky/slp3/
15. Liu, B.: Sentiment analysis and opinion mining. Synth. Lect. Hum. Lang. Technol. **5**(1), 1–167 (2012)
16. McTear, M.F.: Spoken dialogue technology: enabling the conversational user interface. ACM Comput. Surv. (CSUR) **34**(1), 90–169 (2002)
17. Mesnil, G., Dauphin, Y., Yao, K., Bengio, Y., Deng, L., Hakkani-Tur, D., He, X., Heck, L., Tur, G., Yu, D., Zweig, G.: Using recurrent neural networks for slot filling in spoken language understanding. IEEE/ACM Trans. Audio, Speech, Lang. Process. **23**(3), 530–539 (2015)
18. Murray, I.R., Arnott, J.L.: Toward the simulation of emotion in synthetic speech: a review of the literature on human vocal emotion. J. Acoust. Soc. Am. **93**(2), 1097–1108 (1993)
19. Murray, I.R., Arnott, J.L.: Implementation and testing of a system for producing emotion-by-rule in synthetic speech. Speech Commun. **16**(4), 369–390 (1995)
20. Narayanan, S., Georgiou, P.G.: Behavioral signal processing: deriving human behavioral informatics from speech and language. Proc. IEEE **101**(5), 1203–1233 (2013)
21. Paiva, A.: Affective Interactions: Toward a New Generation of Computer Interfaces?. Springer, Berlin (2000)
22. Pang, B., Lee, L., Vaithyanathan, S.: Thumbs up?: sentiment classification using machine learning techniques. In: Proceedings of the ACL-02 Conference on Empirical Methods in Natural Language Processing-Volume 10, pp. 79–86. ACL (2002)
23. Pennebaker, J.W., Graybeal, A.: Patterns of natural language use: disclosure, personality, and social integration. Curr. Dir. Psychol. Sci. **10**(3), 90–93 (2001)
24. Pentland, A.S.: Social signal processing [exploratory dsp]. IEEE Signal Process. Mag. **24**(4), 108–111 (2007)
25. Picard, R.W., Picard, R.: Affective Computing, vol. 252. MIT Press, Cambridge (1997)
26. Ritchie, G.: Current directions in computational humour. Artif. Intell. Rev. **16**(2), 119–135 (2001)
27. Riva, G.: The sociocognitive psychology of computer-mediated communication: the present and future of technology-based interactions. Cyberpsychology Behav. **5**(6), 581–598 (2002)
28. Scherer, K.R., Banse, R., Wallbott, H.G., Goldbeck, T.: Vocal cues in emotion encoding and decoding. Motiv. Emot. **15**(2), 123–148 (1991)
29. Schröder, M., Bevacqua, E., Cowie, R., Eyben, F., Gunes, H., Heylen, D., Maat, M.T., McKeown, G., Pammi, S., Pantic, M., Pelachaud, C., Schuller, B., de Sevin, E., Valstar, M., Wöllmer, M.: Building autonomous sensitive artificial listeners. IEEE Trans. Affect. Comput. **3**(2), 165–183 (2012)
30. Schuller, B.: Towards intuitive speech interaction by the integration of emotional aspects. In: Proceedings IEEE International Conference on Systems, Man and Cybernetics, vol. 6. IEEE, Yasmine Hammamet, Tunisia (2002). 6 pages
31. Schuller, B., Batliner, A.: Computational Paralinguistics: Emotion, Affect and Personality in Speech and Language Processing. Wiley, Hoboken (2013)
32. Schuller, B., Mousa, A.E.D., Vasileios, V.: Sentiment analysis and opinion mining: on optimal parameters and performances. WIREs Data Min. Knowl. Discov. **5**, 255–263 (2015)
33. Serdyuk, D., Wang, Y., Fuegen, C., Kumar, A., Liu, B., Bengio, Y.: Towards end-to-end spoken language understanding. In: 2018 IEEE International Conference on Acoustics, Speech and Signal Processing (ICASSP), pp. 5754–5758. IEEE (2018)

34. Shriberg, E., Stolcke, A.: Prosody modeling for automatic speech recognition and understanding. In: Mathematical Foundations of Speech and Language Processing – The IMA Volumes in Mathematics and its Applications, vol. 138, pp. 105–114. Springer, New York (2004)
35. Strapparava, C., Stock, O., Mihalcea, R.: Computational humour. In: Emotion-oriented Systems, pp. 609–634. Springer, Berlin (2011)
36. Torfi, A., Shirvani, R.A., Keneshloo, Y., Tavvaf, N., Fox, E.A.: Natural language processing advancements by deep learning: a survey (2020). arXiv:2003.01200
37. Tur, G., Celikyilmaz, A., He, X., Hakkani-Tür, D., Deng, L.: Deep learning in conversational language understanding. Deep Learning in Natural Language Processing, pp. 23–48. Springer, Berlin (2018)
38. Tur, G., De Mori, R.: Spoken Language Understanding: Systems for Extracting Semantic Information from Speech. Wiley, Hoboken (2011)
39. Turney, P.D.: Thumbs up or thumbs down?: semantic orientation applied to unsupervised classification of reviews. In: Proceedings of the 40th Annual Meeting on Association for Computational Linguistics, pp. 417–424. ACL (2002)
40. Vaswani, A., Shazeer, N., Parmar, N., Uszkoreit, J., Jones, L., Gomez, A., Kaiser, L., Polosukhin, I.: Attention is all you need (2017). arXiv:1706.03762
41. Vinciarelli, A., Pantic, M., Bourlard, H., Pentland, A.: Social signal processing: state-of-the-art and future perspectives of an emerging domain. In: Proceedings of the 16th ACM International Conference on Multimedia, pp. 1061–1070. ACM (2008)
42. Zhou, H., Huang, M., Zhang, T., Zhu, X., Liu, B.: Emotional chatting machine: emotional conversation generation with internal and external memory. In: Thirty-Second AAAI Conference on Artificial Intelligence (2018)

Printed in the United States
by Baker & Taylor Publisher Services